陶瓷外语系列教材

Ceramic English: A Coursebook on Reading and Translation

陶瓷英语阅读与翻译教程

主　编◎侯晓华
副主编◎邱　慧　邹　丹　余嘉琪

江西高校出版社
JIANGXI UNIVERSITIES AND COLLEGES PRESS
南昌

图书在版编目(CIP)数据

陶瓷英语阅读与翻译教程／侯晓华主编；邱慧，邹月，余嘉琪副主编. －－南昌：江西高校出版社，2024.12. －－（陶瓷外语系列教材）. －－ ISBN 978－7－5762－5326－9

Ⅰ. TQ174

中国国家版本馆 CIP 数据核字第 2024AK6812 号

策划编辑	陈永林	责任编辑	杨良琼	
封面设计	王煜宣	责任印制	李香娇	

出版发行	江西高校出版社
社　　址	江西省南昌市洪都北大道 96 号
邮政编码	330046
总编室电话	0791－88504319
销售电话	0791－88511423
网　　址	www.juacp.com
印　　刷	江西新华印刷发展集团有限公司
经　　销	全国新华书店
开　　本	700 mm×1000 mm　1/16
印　　张	9.25
字　　数	172 千字
版　　次	2024 年 12 月第 1 版
印　　次	2024 年 12 月第 1 次印刷
书　　号	ISBN 978－7－5762－5326－9
定　　价	68.00 元

赣版权登字 －07－2024－1055

版权所有　侵权必究

图书若有印装问题，请随时联系本社印制部（0791－88513257）退换

总序

  2019年5月，习近平总书记在视察江西时，做出了"建好景德镇国家陶瓷文化传承创新试验区，打造对外文化交流新平台"的重要指示。

  作为中国唯一的陶瓷高等学府，景德镇陶瓷大学始终以振兴中国陶瓷工业、弘扬中国陶瓷文化为己任，经过百余年的发展，现已成为全国乃至世界陶瓷人才培养、陶瓷科技创新和陶瓷文化艺术交流的重要基地。为贯彻落实习近平总书记的重要指示，我校外国语学院组织骨干教师编写了"陶瓷外语系列教材"，涵盖英语、日语、法语、德语等语种，内容涉及陶瓷文化、历史、科技、艺术、商务等方面，旨在培养既谙熟陶瓷专业知识，又具备跨语种传播能力的国际化陶瓷人才，用陶瓷外语讲好中国故事，打造对外文化交流新平台，为推动当代中外文化交流贡献景德镇陶瓷大学的绵薄之力。

景德镇陶瓷大学副校长

前言

　　习近平总书记在党的二十大报告中指出要推进文化自信自强，铸就社会主义文化新辉煌。总书记强调，要"增强中华文明传播力影响力。坚守中华文化立场，提炼展示中华文明的精神标识和文化精髓，加快构建中国话语和中国叙事体系，讲好中国故事、传播好中国声音，展现可信、可爱、可敬的中国形象。加强国际传播能力建设，全面提升国际传播效能，形成同我国综合国力和国际地位相匹配的国际话语权。深化文明交流互鉴，推动中华文化更好走向世界"。

　　2023年10月11日，习近平总书记来到景德镇市考察调研。在陶阳里历史文化街区，习近平总书记先后走进南麓遗址、陶瓷博物馆、明清窑作群，同非遗传承人亲切交流，了解制瓷技艺流程、陶瓷文化传承创新和对外交流等情况。习近平总书记指出，中华优秀传统文化自古至今从未断流，陶瓷是中华瑰宝，是中华文明的重要名片。

　　本教材面向英语专业本科生和翻译专业硕士，以中国陶瓷艺术文化为载体，旨在培养学生扎实的英语阅读和翻译技能，进而培养学生的中国陶瓷文化国际传播力，响应习近平总书记在党的二十大报告中提出的"增强中华文明传播力影响力"的号召。

  教材编者精选了不同时代、不同英语国家、不同风格的英文陶瓷文献,全方位地讲述中国陶瓷文化艺术。不同的时代体现了中国陶瓷艺术的发展变化,不同的国家则体现了中国陶瓷文化艺术在国际上传播的广度,不同风格的文章则为学生提供了学习中国陶瓷文化艺术的不同视角。

  教材总共八个单元,按照时间顺序编排,围绕某一朝代或时期的典型代表陶器或瓷器种类展开主题。每个单元主要由三部分构成。第一部分是单元主题内容介绍和主题阅读文章,通过文章阅读培养学生的英语语言技能。第二部分是阅读理解帮助,主要包括文章中涉及的中国传播陶瓷文化和历史相关的背景介绍和陶瓷专业术语的介绍。这部分是本教材的特色,即课程思政内容体现板块。通过详细介绍中国陶瓷文化、历史、艺术等相关知识,弘扬中华文明,树立学生的文化自信,进而促进中国陶瓷文化国际传播。第三部分是课后练习,由阅读能力检测题和翻译技能检测题构成,通过实践训练增强学生的阅读和翻译技能。

  本教材由侯晓华负责编写第一、第二、第三、第六、第七单元内容,邱慧负责编写第四、第五两个单元的内容,邹丹负责编写第八单元的内容,余嘉琪负责审校工作。

  本教材的编者均是第一次编写教材,因而书中错误难免,欢迎各位读者予以批评指正。

编 者
2024 年 8 月 1 日于景德镇陶瓷大学

目录 CONTENTS

Unit 1　Introduction to Ceramics　/001
　Text　Introduction to the Origin of Porcelain　/002
　Aids to Comprehension　/009
　Exercises　/014

Unit 2　Pottery Before the Qin and Han Dynasties　/016
　Text　Pottery Making Before the Qin and Han Periods　/017
　Aids to Comprehension　/026
　Exercises　/030

Unit 3　Three-color Glazed Wares of the Tang Dynasty　/032
　Text　Three-color Glazed Wares of the Tang Dynasty　/033
　Aids to Comprehension　/040
　Exercises　/044

Unit 4　Song Dynasty Kilns in Northern China　/046
　Text　Ting Kilns　/047
　　　　Ru Kilns　/049
　　　　The Cizhou Kilns　/052
　　　　Chun Yao　/055

Aids to Comprehension　/061

Exercises　/064

Unit 5　Song Dynasty Kilns in Southern China　/066

　　Text　The Guan Kilns　/067

　　　　　Lung-chuan Yao　/072

　　　　　The Jingdezhen Kilns　/075

　　Aids to Comprehension　/084

　　Exercises　/088

Unit 6　Blue and White of the Yuan and Ming Dynasties　/090

　　Text　The Making of Blue and White in Late Yuan and Ming Dynasties　/091

　　Aids to Comprehension　/098

　　Exercises　/103

Unit 7　Polychrome Glazed Wares in the Qing Dynasty　/105

　　Text　The Polychrome Glazed Wares in the Qing Dynasty　/106

　　Aids to Comprehension　/114

　　Exercises　/118

Unit 8　Introduction to Chinese Export Porcelain　/120

　　Text　Porcelain Made for Exportation　/121

　　Aids to Comprehension　/128

　　Exercises　/133

Glossary　/135

References　/139

Unit 1 Introduction to Ceramics

Ceramics, which includes **pottery** and **porcelain** wares, is the art of "fire and earth" and an important component of the material and spiritual culture of mankind. Chinese ceramics, in particular, has a long and continuous production history which has never suspended in any period and sustainable to the present with a wide repertoire of types and forms. They are produced with superb techniques and are popular merchandises export to various countries and regions with significant influence that command attention and esteem from people in all parts of the world. As far as the origin of porcelain is concerned, scholars have different opinions and evidence to support their arguments.

Text

Introduction to the Origin of Porcelain

Stephen W. Bushell

1   PORCELAIN was invented in China. The exact date of the invention, however, is wrapped in mystery; it is, in fact, hardly likely that it will ever be definitely settled, as it must have been by a gradual progress in the selection of materials, and in the perfection of processes of manufacture, that porcelain was at last evolved from ordinary **pottery**. For the creation of a scientific classification of ceramic products, we are indebted to **M. Brongniart**, and it will be well first to define the distinctive characteristics of porcelain. Porcelain ought to have a white, **translucent**, hard **paste**, not to be scratched by steel, homogeneous, resonant, completely **vitrified**, and exhibiting, when broken, a conchoidal fracture of fine **grain** and brilliant aspect. These qualities, inherent in porcelain, make it impermeable to water, and enable it to resist the action of frost even when uncoated with **glaze**. These characteristics of the paste, especially the translucence and vitrification, define porcelain very well. If either of these two qualities be wanting, we have before us another kind of pottery; if the paste possesses all the other properties, with the exception of translucence, it is a **stoneware**; if the paste be not vitrified, it belongs to the category of **terra cottas** or of **faience**.

2   The Chinese define porcelain, which they call 瓷 (ci) as a hard, compact, fine-grained pottery 陶 (tao), and distinguish it by the clear, resonant note which it gives out on percussion, and by the fact that it can not be scratched by a knife. They do not lay so much stress on the whiteness of the paste, nor on its translucency, so that some of the pieces may fail in these two points, when the fabric is coarse; and yet it would be difficult to separate them from the porcelain class. The paste of the ordinary ware, even at ***Ching-te-chen***, is composed of more heterogeneous materials than that fabricated in European factories, and may even be reduced in some cases to a mere layer of true porcelain earths plastered over a substratum of yellowish gray clay. The Chinese separate, on the other hand, dark-colored stonewares, like the reddish-

yellow ware made at **Yi-hsing**, in the province of **Kiangnan**, known to us by the Portuguese name of **boccaro**, or the brown stoneware produced at **Yang-chiang**, in the southern part of the province of **Kwang-tung**, which is coated with colored enamels, and is often put in European collections among the monochrome porcelains. The Chinese word for pottery in its widest sense is *tao*, which includes all ceramic products, from common **earthenware** to porcelain. Like many of the great nations of antiquity, they claim for themselves the invention of the **potter's wheel**. M. Brongniart is inclined to admit their claim, and even attempts to trace the route by which it may have reached Egypt, through Scythia and Bactria; but such speculations seem too hazardous. It was certainly known to the Egyptians at a very early period, probably not later than twenty-five hundred years before **our era**. Scenes depicted at **Beni Hassan** and at **Thebes** show us the Egyptian potters at work, and figure the simple wheel, consisting of a flat disk or hexagonal table, placed on a stand, which appears to have been turned with the left hand while the vase was shaped with the right. The Chinese claims go back to about the same period, as they attribute the invention of the potter's wheel to the director of pottery attached to the court of the fabulous Emperor **Huang Ti**, to whose reign they carry back their cyclical system of chronology, starting from a date corresponding to 2637 B. C.. The **Emperor Shun**, whose reign is placed in 2255 – 2206 B. C., is generally credited with the first improvements in the art of welding clay. **Ssu-ma Ch'ien**, the **Herodotus** of China, the compiler of the *Shi Ji*, the first of the dynastic histories, says in his biography of Shun, that before he came to the throne, he made pottery at **Ho-pin**. **Père d'Entrecolles** describes the immense value a Chinaman attaches to any pieces of pottery he can attribute to the reigns of Yao and Shun. Tradition says that Yao adored simplicity, and had his sacrificial vessels fashioned of plain yellow earthenware and that Shun was the first to have them glazed, and the credulous collector classifies his prehistoric pieces accordingly. Coming to more historical times, the period of the **Chou dynasty** (1046 – 256 B. C.), the third of the **Three Ancient dynasties**, its founder, **Wu Wang**, is recorded to have sought out a lineal descendant of the Emperor Shun, on account especially of his hereditary skill in the manufacture of pottery, to have given him his eldest daughter in marriage, and to have appointed him feudal ruler of the state of **Ch'en** (now Ch'en-chou Fu, in the province of Honan), to keep up there the ancestral worship of his accomplished ancestor.

3 In **K'ao kung chui**, an artificer's manual of the same period, there is a short section on pottery, which gives the names and measurements of several kinds of cooking vessels, sacrificial vases, and dishes, in the fabrication of which the different processes of fashioning upon the wheel and of molding are clearly distinguished. The vessels are described as having been made by two classes of workmen, called respectively t'ao-jén, "potters," and fang-jén, "molders."

4 But few specimens of pottery that can be certainly referred to the Three Ancient dynasties have survived to the present day, although ritual vessels and other antiques of bronze are to be seen in native collections by thousands. These last often have inscriptions upon them, beginning perhaps with the number of the month, the waxing or waning period of the moon, the day of the month and its cyclical number; rarely is the year of the reigning sovereign or feudal suzerain prefixed; never his name, as far as I know. It was during the Han dynasty, which reigned from 202 B. C. to 220 A. D., that the system of dividing the reigns into periods of years with honorific titles (nian hao) was inaugurated in 163 B. C.. This provided for the first time a convenient means of dating vases and other objects.

5 Bricks and tiles are among the most useful of ceramic products. They may even rank as historical monuments when inscribed. The Chinese antiquary collects them in chronological series to show the changes in the style of the written character, or puts one upon his writing-table for daily use, excavated into the shape of an ink pallet. They were first molded, with the date inscribed on one side, during the Han dynasty. Some of the pottery of the period is also inscribed.

6 With regard to the origin of porcelain in China, the Chinese themselves confess that previous to the commencement of the Tang dynasty, in 618 A. D., there are no criteria for forming an opinion. The names of some score of different sacrificial vases, drinking vessels, and other objects may be collected from books, but nothing

is said about their structure or place of production. It was reserved for a western scholar to carry back the invention to the Han dynasty, and to date it precisely as between 185 B. C. and 87 A. D. . These dates, adopted by **M. Julien** in the preface of one of his book, have been generally followed by writers on the subject, as derived from Chinese records, although based, as we shall show, on fallacious grounds. They are deduced from a short note in the appendix to the memoir on the administration of porcelain in the annals of **Fou-liang**, which reads, "The ceramic manufacture of **Hsin-p'ing** according to local tradition, was founded in the time of the Han dynasty, and was probably of strong, heavy, and roughly finished material, **moulded** and **fashioned** after methods handed down from ancient times."

7   The Chinese names of the geographical dictionaries from which these facts are taken are given in footnotes, but all the trouble of reference would have been saved had M. Julien known that Hsin-p'ing was the original name of Fou-liang Hsien. It is recorded in the geographical section of the official annals of the Tang dynasty (*Tang Shu*) that this walled city was founded under the name of Hsin-ping, in the fourth year of the period Wude (621 A. D.), with jurisdiction over a tract which formed part of the old district of Po-yang; that it was re-established in the fourth year of Kaiyuan (716 A. D.), under the new name of Hsin-ch'ang; and that its name was finally changed to Fou-liang (which it has kept to the present day) in the first year of the period Tianbao (742 A. D.).

8   Hsin-ping occurs constantly in different pages of the annals quoted above as the old name of Fou-liang, and it is, besides, referred to more than once in the last three books of the ***Ching-te-chen T'ao Lu***, which are omitted in Julien's translation. An extract, for example, is quoted in the book, from the biography of **Chu Sui**, styled Yu-héng, an official under the Tang dynasty, who was superintendent at Hsin-p'ing, when, in the first year of the period Jinglong (707 A. D.), an imperial decree was received by the Governor of **Hung-chou**, ordering him to supply with all speed a number of sacrificial utensils for the imperial tombs. Chu Sui is described as having pushed on the work so energetically that they were all sent before the end of the year.

9 Hung-chou is the old name of the modern Nan-ch'ang Fu, the chief city of the province of Kiangsi, and **Jao-chou**, within the bounds of which lies Fou-liang Hsien, is stated in the *Annals of the Tang Dynasty* to have been actually at that time under the jurisdiction of the Governor of Hung-chou.

Text Related Information

This text is adapted from chapter one of the book *Oriental Ceramic Art* (1897), written by Stephen W. Bushell, published by D. Appleton and Company, New York.

Selected Plates Related to the Text

Plate 1 Wine jar with underglaze blue decoration. Height: 34 cm, max. diameter: 36 cm. Ming dynasty.

Unit 1　Introduction to Ceramics

Plate 2　Celadon jar with high-relief decoration. Height: 33.7 cm. Yuan dynasty.

Plate 3　Porcelain tile with fahua decoration. Length: 21 cm, width: 23 cm. Ming dynasty.

Plate 4　Vase painted in iron-black on a cream slip under a turquoise glaze. Height: 28 cm. Ming dynasty.

Plate 5　Square Cizhou-type stoneware wine flask, slipped, and painted in iron brown. Height: 31.5 cm, width: 36 cm, depth: 11.2 cm. Yuan to early Ming dynasty.

Aids to Comprehension

Notes

(1) Stephen W. Bushell: Stephen Wooton Bushell 斯蒂芬·伍敦·卜士礼（1844—1908），又译斯蒂芬·伍敦·波西尔、斯蒂芬·伍敦·布绍尔，英国人，出生于英国肯特郡，毕业于伦敦大学盖伊医学院（Guy's Hospital Medical School），1866年毕业后在盖伊医院任住院外科医生，次年在贝特莱姆皇家医院（Bethlem Royal Hospital）担任驻院医生（resident medical officer），1868年获伦敦大学医学博士学位。1868年1月，卜士礼在雒魏林的推荐下，前往北京担任英国驻华使馆医师，并兼任京师同文馆医学教习。他在中国居住长达32年，不仅学会了中文，还撰写了许多关于中国艺术、钱币学、地理、历史等方面的论文。1900年退休后他回到英国，此后出版了《中国美术》（Chinese Art，1905—1906）、《中国瓷器》（Chinese Porcelain，1908）、《中国陶瓷图说》（Description of Chinese Pottery and Porcelain，1910）等著作。1908年在英国米德赛克斯（Middlesex）逝世。

(2) Ching-te-chen 景德镇市，位于江西省东北部与安徽省的交界处，素有"瓷都"之称。与河南朱仙镇、湖北汉口镇、广东佛山镇并称"四大名镇"。东晋时此地设新平镇，唐武德年间就镇设县，称新平县，后并入鄱阳。开元时重新置县，改称新昌县，于昌江之北另设治所。天宝元年更名为浮梁县，后于1960年撤销。景德镇在宋代以前称新平镇，因在昌江南岸，又称昌南镇。景德年间（1004—1007年），因该镇所造瓷器名扬天下，被赐名为景德镇。

(3) Yi-hsing 宜兴

(4) Kiangnan 江南。这里指江南的江苏省。

(5) Yang-chiang 阳江，位于广东省，这里有著名的石湾窑。宋代有关阳江石湾窑的部分文献记载：广东瓷器，宋代以阳春、阳江为最著。此说见于清雍正时所修的《广东通志》，中外学者多有援述。如英国人卜士礼在其著作《中国美术》一书中赞道："阳春之南有县曰阳江，与阳春同隶于肇庆，而较阳春距海尤近，乃广窑中最著名之器也。其窑质致密坚固，极耐磨折……

昔曾有此窑所制之宝蓝瓶,自北京宫中发出(清乾隆时)交唐英(当时的督陶官)在官窑中依式仿造,其精巧可知也。"阳江窑最初的烧造年代有不同的说法,据晚清许之衡在《饮流斋说瓷》一书中说:"广窑,宋南渡后所建,在广东肇庆阳江,胎质粗而色褐(即灰色),所制器多作天蓝色,惟不甚匀耳。"

(6) Kwang-tung 广东省

(7) our era 我们所处的时代,这里指19世纪末。

(8) Beni Hassan 埃及贝尼哈桑考古遗址。该遗址位于尼罗河东岸,在开罗以南大约245千米处。该遗址以石凿墓穴著称,墓穴都是第十一王朝和第十二王朝期间上埃及第十六州即奥里克斯州高级官吏的埋葬之处。

(9) Thebes 底比斯城。古希腊城市,与雅典、斯巴达并称为希腊三大主要城邦。

(10) Huang Ti 黄帝,又称轩辕氏,是中国古代传说中一位伟大的君主和英雄。他的一生充满了传奇色彩——他不仅统一了部落,还创造了许多文化和科技的奇迹。

(11) Emperor Shun 舜帝,传说中父系氏族社会后期部落联盟领袖,姚姓,一作妫姓,名重华,"三皇五帝"之一。舜为东夷族群的代表,生有重瞳,孝顺友爱,善于制陶。舜得到四岳推荐,经过重重考验,得到尧的认可与禅位而称帝于天下,其国号为"有虞",故号为"有虞氏帝舜"。帝舜、大舜、虞帝舜、舜帝皆虞舜之帝王号,故后世以舜简称之。舜曾在河滨地区从事制陶工作。他改善了陶器的质量,并推动了陶器工艺的发展。

(12) Ssu-ma Ch'ien 司马迁(前145—前90),字子长,夏阳(今陕西韩城南)人。中国西汉伟大的史学家、文学家、思想家,任太史令。著有《史记》《悲士不遇赋》等。

(13) Herodotus 希罗多德(约前484—前425),古希腊作家、历史学家。他把旅行中的所闻所见以及第一波斯帝国的历史记录下来,著成《历史》一书,希罗多德因此被尊称为"历史之父"。

(14) Ho-pin 河滨。这里指"舜陶河滨"的传说。河滨即现在山东省菏泽市。

(15) Père d'Entrecolles: Père Francois Xavier d'Entrecolles 殷弘绪(1664—1741),法国传教士。他曾在中国景德镇居住过七年,1712年他写信给法国的传教士,详细地介绍了瓷器的原材料和制作方法,从而使法国人在法国本地仿造出瓷器。

（16）Chou dynasty　周朝

（17）Three Ancient dynasties　指夏、商、周三个朝代。

（18）Wu Wang　周武王

（19）Ch'en　陈国。妫满是舜帝的嫡系子孙,周武王将自己的长女大姬嫁给了他,并把他封为侯爵,封地在陈,建立陈国,负责供奉祭祀舜帝。

（20）*K'ao kung chui*　《考工记》,出于《周礼》,是中国春秋战国时期记述官营手工业各工种规范和制造工艺的文献。书中记述了木工、金工、皮革、染色、刮磨、陶瓷6大类30个工种的内容,体现了当时中国的科技及工艺水平,此外其中还有数学、地理学、力学、声学、建筑学等多方面的知识和经验总结。《考工记》中将陶瓷工种分为陶人和瓬人。

（21）Chou dynasty　周朝,始建于公元前1046年,灭亡于公元前256年。周朝是中国历史上继商朝之后的第三个华夏族奴隶制王朝。周朝分为西周(前1046—前771年)和东周(前770年—前256年)两个时期。

（22）M. Julien: Stanislas Aignan Julien　儒莲(1797—1873),法国汉学家,法兰西学院院士。他对中国语言、文化和社会有着广泛而深入的了解,译著成果颇丰。经过对中文典籍的潜心钻研,他分别于1840年和1856年将《天工开物》和《景德镇陶录》翻译为法文,向欧洲介绍中国陶瓷制作技术。

（23）Fou-liang　浮梁县,隶属于江西省景德镇市,位于江西省东北部、赣皖二省交界处。唐天宝元年(742年),更名为浮梁。

（24）Hsin-p'ing　新平。景德镇始称"新平镇",又称"昌南镇",唐代时改名"浮梁"。唐开元四年间(716年),新平镇改为新昌县。景德镇制瓷的历史最早可以追溯到汉代,距今已有1800多年的历史。史料记载,"新平治陶,始于汉世"。这一时期,景德镇处于制陶和制作原始瓷器或早期瓷器的阶段。到了唐代,景德镇的制瓷业有了很大的发展。

（25）Ching-te-chen T'ao Lu　《景德镇陶录》,清代蓝浦著,郑廷桂补辑。

（26）Chu Sui　褚绥,字玉衡,晋州人,唐景隆(707—710年)初为新平(景德镇)司务,奉诏监烧献陵祭器。

（27）Hung-chou　洪州,即今江西省南昌市。

（28）Jao-chou　饶州府,明清行政区划名。明洪武二年(1369年),鄱阳府改为饶州府,府治鄱阳县,领鄱阳(今江西省鄱阳县)、德兴(今江西省德兴市)、安仁(今江西省余江区)三县,余干(今江西省余干县)、浮梁(今江西省浮梁县、景德镇市区)、乐平(今江西省乐平市)三州。民国元年(1912年),饶州府废。

Ceramic Terminology

(1) porcelain: high-fired pottery that, by western standards, is hard, dense, resonant when struck, impervious to liquid, white, and translucent. Porcelain is fired at a temperature in excess of about 1,250 ℃. 按照西方标准,瓷是一种经1250 ℃以上高温烧制的陶器,其胎体坚固、致密、不透水,通常呈白色半透明状,敲击时发出洪亮的响声。

(2) pottery: refers to all objects made from fired clay, whether they are earthenware, stoneware, porcelaneous ware, or porcelain; synonymous with "ceramics" 陶器,即所有用烧制过的黏土制作的东西,无论是陶器、炻器、近似瓷器还是瓷器;其与英文中的"ceramics"一词同义。

(3) translucent: (of a substance) allowing light, but not detailed shapes, to pass through, semitransparent (物质)半透明的

(4) paste: a mixture consisting mainly of clay and water that is used in making ceramic ware, especially a mixture of low plasticity based on kaolin for making porcelain (制作陶、瓷器用的)湿黏土,尤指以高岭土为基础的低塑性黏土。

(5) vitrify: to convert (something) into glass or a glass-like substance, typically by exposure to heat. 玻璃化;使……呈玻璃状

(6) grain 纹理;细颗粒

(7) glaze: in essence, a glassy coating on the surface of a ceramic. A glaze serves the twofold function of helping to seal the clay body and decorating the object. Most glazes are predominantly composed of silica, with other materials—known as fluxes—added to the silica, primarily to lower its melting point. Alumina is almost always added to increase the viscosity of the glaze. Glazes may be applied to a raw body and fired with the body, or they may be applied to a pre-

fired, or biscuit, body and given a separate firing, known as a glost firing. 釉，本质上是陶瓷器物表面的一层玻璃状覆盖物。其功能有二，一是如涂层般保护坯体，二是起装饰作用。大多数釉料主要由二氧化硅和其他材料组成。这些材料便是助熔剂，用来降低熔点。釉料中通常加入氧化铝以增强黏性。坯体施釉后再烧制，或者将釉料施在烧制后的坯体即素坯上然后再二次烧制，此为釉烧。

(8) stoneware: vitrified, high-fired pottery made of clay to which a proportion of other materials may be added to achieve good working and firing properties. Stoneware is fired in excess of about 1,200 ℃. It is dense, hard, resonant when struck, and impervious to liquid; it may be light or dark in color, but it is not translucent. 炻器；黏土搭配适量提高制作和烧制效果的其他材料，经过1200 ℃以上高温烧制而成的陶瓷。其胎体致密、坚固，不透水、不透明，敲击时发出响声，器表颜色或深或浅。

(9) terra cotta: a brownish-red clay that has been baked and is used for making things such as flower pots, small statues, and tiles. 烧制后的红色黏土，用来制作花盆、小型塑像或砖瓦。

(10) faience: earthenware decorated with opaque colored glazes 彩釉装饰的陶器；彩陶

(11) boccaro 宜兴红陶（葡萄牙语）

(12) erthenware: low-fired pottery made from common clay to which a proportion of other materials may be added to achieve good working and firing properties. Earthenware is porous and permeable; in color, it may range from light buff to tan, red, brown, or black, depending on the clay and firing conditions. Earthenwares are usually fired between about 600 ℃ and 1,100 ℃. 普通黏土搭配适量提高制作和烧制效果的其他材料，经过低温烧制而制成的陶器。

(13) potter's wheel 陶车；拉坯车

(14) mould: to form (an object with a particular shape) out of easily manipulated material 用可塑材料塑成（某种形状的物体）；塑造；模

(15) fashion: here used as a verb, to make into a particular or the required form 制作

Exercises

Reading Comprehension Questions

Please answer the following questions according to the text.

(1) What are the characteristics of porcelain according to M. Brongniart?

(2) How is the Chinese definition of porcelain different from M. Brongniart?

(3) What are the features of inscriptions on antiques of bronze of the Three Ancient dynasties?

(4) What is the time of the invention of porcelain according to a western scholar?

(5) Please tell the relationship between Xinping and Jingdezhen.

Translation

A. Please translate the following paragraphs into Chinese, paying special attention to ceramic terminology, context and the social background involved.

(1) Porcelain was invented in China. The exact date of the invention, however, is wrapped in mystery; it is, in fact, hardly likely that it will ever be definitely settled, as it must have been by a gradual progress in the selection of materials, and in the perfection of processes of manufacture, that porcelain was at last evolved from ordinary pottery. For the creation of a scientific classification of ceramic products, we are indebted to M. Brongniart, and it will be well first to define the distinctive characteristics of porcelain. Porcelain ought to have a white, translucent, hard paste, not to be scratched by steel, homogeneous, resonant, completely vitrified, and exhibiting, when broken, a conchoidal fracture of fine grain and brilliant aspect. These qualities, inherent in porcelain, make it impermeable to water, and enable it to resist the action of frost even when uncoated with glaze. These characteristics of the paste, especially the translucence and vitrification, define porcelain very well. If

either of these two qualities be wanting, we have another kind of pottery before us; if the paste possesses all the other properties, with the exception of translucence, it is a stoneware; if the paste be not vitrified, it belongs to the category of terra cottas or of faience.

(2) With regard to the origin of porcelain in China, the Chinese themselves confess that previous to the commencement of the Tang dynasty, in 618 A.D., there are no criteria for forming an opinion. The names of some score of different sacrificial vases, drinking vessels, and other objects may be collected from books, but nothing is said about their structure or place of production. It was reserved for a western scholar to carry back the invention to the Han dynasty, and to date it precisely as between 185 B.C. and 87 A.D.. These dates, adopted by M. Julien in the preface of one of his book, have been generally followed by writers on the subject, as derived from Chinese records, although based, as we shall show, on fallacious grounds. They are deduced from a short note in the appendix to the memoir on the administration of porcelain in the annals of Fou-liang, which reads, "The ceramic manufacture of Hsin-p'ing according to local tradition, was founded in the time of the Han dynasty, and was probably of strong, heavy, and roughly finished material, moulded and fashioned after methods handed down from ancient times."

B. Please translate the following paragraph into English.

瓷器是中国的伟大发明。这是中国人民为人类历史发展所作出的卓越贡献之一。瓷器与陶器不同：瓷器所用原料是瓷土（包括高岭土、长石、石英或含有这些成分的瓷石）。瓷器必须在表面上施一层在高温下烧成透明体或半透明体的釉，要经过1200 ℃以上的高温焙烧。坯体烧结后，会出现一种坚韧的针状结晶作为骨架，针状结晶之间填充玻璃体。因此，坯体没有吸水性或吸水性极弱，叩之可发出清脆的类似金属的声音。

Unit 2　Pottery Before the Qin and Han Dynasties

Pottery, as one of the first necessities of mankind, is among the earliest of human inventions. In a rude form it is found with the implements of the late Stone Age, before there is any evidence of the use of metals, and all attempts to reconstruct the first stages of its discovery are based on conjecture alone. We have no knowledge of a Stone Age in China, but it may be safely assumed that pottery there, as elsewhere, goes back far into prehistoric times. Its invention is ascribed to the mythical **Shen-nung**, the **Triptolemus** of China, who is supposed to have initiated the people in the cultivation of the soil and other necessary arts of life. Huang Ti, the semi-legendary yellow emperor, in whose reign the cyclical system of chronology began (2697 B. C.), is said to have appointed "a superintendent of pottery, **K'un-wu**, who made pottery," and it was a commonplace in the oldest Chinese literature that the great and good emperor *Yu Ti Shun* (2317 – 2208 B. C.) "highly esteemed pottery." Indeed, the Han historian *Ssu-ma Ch'ien* (163 – 85 B. C.) assures us that Shun himself, before ascending the throne, "fashioned pottery at Ho-pin," and, needless to say, the vessels made at Ho-pin were "without flaw."

Text
Pottery Making Before the Qin and Han Periods

Suzanne G. Valenstein

1 According to ancient literary sources, China's first hereditary dynasty, the Xia, came into power in the general area of the middle **Yellow River** valley about the twenty-first century B. C. While this legendary **Xia dynasty** still remains something of an enigma, the theory that it can be identified with what is known as the **Erlitou culture** has gained wide acceptance in recent years. This bronze-producing Erlitou culture is typified by the **remains** found at the Erlitou site in *Yanshi xian*, in northwestern Henan Province. Similar cultural remains have been found in other parts of the province, as well as in southern **Shanxi**, eastern Hubei, eastern **Shaanxi**, and southern Hebei provinces. Stratigraphic evidence places the Erlitou culture between the late Henan **Longshan culture** and the **Zhengzhou phase** of the Shang dynasty, and **carbon-14** tests have shown that the Erlitou site itself is datable to about 1900 – 1600 B. C.

2 Erlitou ceramics seem to fit quite comfortably into the segment of time between the late Henan Longshan and early Shang period, showing an evolution from the later Longshan pottery to the early Shang **material. Excavated** Erlitou/Xia pottery has mainly consisted of natural clay or sand-tempered gray earthenwares. Black pottery—including **black-slipped pottery**—brownish, red, and white wares have been seen less frequently.

3 **Geometric-impressed pottery**. In the **Shang dynasty**, a very large and widely divergent family of ceramics with a variety of impressed geometric patterns had begun to appear. In classifying these ceramics, Chinese scholars rank them as **hard earthenwares**, noting that the brownish-toned bodies can, in fact, range from fairly soft to what would be described as stoneware in the West. Occasionally these vessels have a surface luster resembling a thin layer of glaze. Excavations of Shang-and-Zhou period sites have frequently unearthed this geometric-impressed pottery in company with contemporary glazed wares.

4　Geometric-impressed ceramics have been found in sites along the middle and lower Yellow River valley, the middle and lower **Yangtze River** valley, and the Southeast China coast. The shapes and decorations have varied somewhat from region to region. The preponderance of these wares was manufactured in Southeast China, where both the types of vessels and kinds of decoration have been more numerous than in the north.

5　**"Cord-marked" pottery**. Common "cord-marked" wares for everyday use, a tradition continued from the Longshan cultures, might be seen as the staple of Shang ceramics. Many of these vessels show evidence of having been constructed in a cord-lined **mold** or by the **beater-and-pad technique**; mouths and feet have frequently been shaped on a wheel and have often been nicely **finished**. The bodies are either untempered or sand-tempered common clay, ranging from relatively fine to very coarse in texture; they may be gray or, less frequently, red or brown. Two *li* **tripods** (Plate 6), one gray and the other with a coarse, sand-tempered, red body, are reliably reported to have been found in Shandong Province. They are honest, straightforward, utilitarian "cord-marked" pots that were designed to expose a large surface to the cooking fire.

6　**Shang lacquered pottery**. Excavations in 1975 at the **Xiaotun** site near Anyang, in Henan Province, produced fragments of lacquer-painted, black-surfaced pottery. Some of these fragments had red-lacquer designs taken from Shang-dynasty bronze vessels. While lacquered pottery was never very popular in China, its manufacture did continue at least through the Han dynasty.

7　**Architectural pottery**. The earliest type of architectural pottery manufactured by the Chinese seems to have been pipes to carry water. Remains of such earthenware water pipes have been found at both the **Erlitou** and **Zhengzhou sites** in Henan Province.

8　Western Zhou pottery provides a logical place to digress for a moment and discuss an important fact about the development of all Chinese pottery. Unlike automobiles from a Detroit assembly line, where this year's model is distinct from last year's or next year's car, Chinese ceramics exhibit a gradual, albeit constant, evolution in both style and physical characteristics, with relatively few major innovations interrupting this slow and measured pace. Throughout Chinese history, while the temper of the time certainly made itself felt in all creative efforts, changes in taste generally occurred by degrees. As might be expected from the empirical methods of manufacture that were used, the material qualities of fabricated objects altered progressively as well. It is therefore axiomatic that Chinese pottery—which on the whole developed in barely perceptible successive stages—usually cannot be pigeonholed in precise or narrow dates, either by Chinese or western calendars.

9 **Unglazed pottery**. Unglazed pottery produced after the Zhou conquered the Shang illustrates the above point well. It shows little dramatic change; instead, it evolved predictably from the typical Shang style to one characteristic of the Zhou. In fact, it is often somewhat difficult to state with certainty whether a given piece dates to the late Shang or early Zhou. A case in point is the unburnished, **high-footed bowl** (Plate 7) with carved and **incised** decoration that was illustrated in the first edition of this book as an example of late Shang-dynasty gray wares coping the shape of bronze serving vessels. Recent archaeological reports have shown rather convincingly that this vessel should be attributed to the Western Zhou period instead.

10 Ordinary Western Zhou unglazed wares still were the relatively crude, gray or red "cord-marked" variety stemming from earlier periods, and, as before, many examples of this long-enduring type show marks from the mold or beater. As demonstrated above, the gray earthenwares of better quality that were produced during the Shang dynasty continued to be made in the early Zhou era and, like the Shang wares, this finer Zhou gray pottery frequently shows strong bronze influences in shapes and designs.

11 **Glazed pottery**. A furthering of the tradition of high-fired glazes that was inherited from the Shang era can be seen in the considerable amount of glazed pottery excavated in recent years from Western Zhou-period sites in at least ten provinces. It is obvious that ceramic technology had advanced, for some of these wares correspond in every respect to what is known as **stoneware** in the West. An interesting group of ceramics was excavated in 1959 from two Western Zhou-period tombs in the suburbs

of **Tunxi** in Anhui Province. Among the seventy one glazed vessels that were found, some were fairly low fired, with porous bodies and poorly fitting brownish glazes, while others were higher fired and had much harder bodies and well-fitting, grayish green glazes.

12 **Geometric-impressed pottery**. The tradition of ceramics with geometric designs **stamped** into the surface while the clay was still damp continued in this period, although such wares have seldom been found in Yellow River valley Western Zhou sites. On the other hand, excavations of contemporary sites farther south, in Jiangxi, Jiangsu, and Zhejiang provinces, have unearthed large quantities of this material; its manufacture can be documented in Southeast China down to Guangdong Province.

13 **Architectural pottery**. Both flat and cylindrical roofing tiles were in use by the Western Zhou era, and pottery rings for the construction of wells were being manufactured.

14 **Pottery** manufactured early in **the Spring and Autumn era** is scarcely distinguishable from that of the late Western Zhou, but as time went on, shapes gradually changed, and new types of vessels were introduced.

15 **Unglazed pottery**. Unglazed ceramics comprise the typical, fairly coarse, "cord-marked" utility products and the finer gray earthenwares that, like their Western Zhou forerunners, frequently show the influence of bronzes. Occasionally the surface of these superior gray wares had been **polished** with a hard instrument while the clay was in what is known as a leather-hard state. This polishing produced burnished designs that contrast with the duller, dark-surfaced body. A softer-bodied gray earthenware also came into being at this time. It is assumed that these lower-fired ceramics, which would have been less expensive to produce than other wares, were used exclusively as **mortuary pottery**.

16 **Painted unglazed pottery**. Two tombs belonging to members of the ruling family of the state of **Ju**, of the late Spring and Autumn period, were excavated near **Junan xian**, Shandong Province, in 1975. Among the tomb furnishings were a number of gray-bodied earthenwares that had been dressed with black **slip**, over

which various designs had been painted in red pigment. While this type of material was somewhat rare in the Spring and Autumn era, painted gray pottery would later become quite popular.

17 **Glazed pottery**. During the Spring and Autumn period, the quality of high-fired glazed pottery continued to improve; some objects now show evidence that a wheel was used in their construction. Bodies are fairly fine and hard; they can be grayish white, yellowish white, or dark brown. The green glazes frequently lean toward a yellow or gray tone. By the late Spring and Autumn period, production of high-fired, glazed near-stoneware and stoneware was very rare in the Yellow River valley. Manufacture was almost entirely concentrated in the area south of the Yangtze River, most notably in southern Jiangsu and northern Zhejiang provinces. This region was to hold a dominant position in the production of high-fired glazed ceramics for centuries to come.

18 **Geometric-impressed pottery**. Like their high-fired glazed counterparts—with which they have frequently been excavated—Spring and Autumn ceramics with impressed geometric designs developed from similar Western Zhou pottery. These hard-bodied wares are relatively uncommon in the Yellow River valley; they were produced for the most part in the middle and lower Yangtze River valley and along the Southeast China coast.

19 During **the Warring States Era**, pottery making made progresses.

20 **Burnished pottery**. The shapes of burnished gray wares with blackish surface were often copied from those made of other materials. Designs, which are slightly glossy and a little darker than the matte ground, were derived from those on contemporary bronzes and lacquers. While this technique of burnishing decorative figures on pottery originated in the Spring and Autumn era, it became especially popular during the Warring States period, when some extraordinary examples were produced. Burnished designs could be combined with patterns that were incised into the body before it was completely dry; and again, late Zhou gray wares might just be enriched with bronze-derived incised decoration. Such burnished pottery can be represented here by Plate 8.

21 **Lacquered pottery**. The technique of lacquering pottery, which originated in the Shang period, was also followed to some extent in the Western and Eastern Zhou eras. While the method of application seems to have been fairly successful, and some known examples are in reasonably good condition even today, lacquered pottery apparently was never as popular as the contemporary lacquered wood.

22 **"Cord-marked" pottery**. The serviceable, workaday "cord-marked" family of ceramics had done yeoman duty since earliest times. It is still apt to be found in shapes more natural to clay and seldom shows evidence of the influence of bronzes. These "cord-marked" gray or red earthenwares, which, as always, have bodies that range from fine to coarse in **texture**, are represented here by two gray jars (Plate 9) that are reported on good authority to have come from Shandong Province. They certainly are not grand pieces, but they have the honest design of a type of utensil that has served well for many centuries.

23 **Geometric-impressed pottery**. The tradition of pottery with impressed geometric patterns persisted in southern China from a period contemporary with the Shang dynasty through the early Han dynasty. This tradition can be seen in numerous Warring States unglazed earthenwares and stonewares excavated from Jiangsu to Guangxi provinces. In the vicinity of Jiangsu and Zhejiang provinces, however, production of geometric-impressed material, like that of glazed stonewares, seems to have suffered considerably during the late Warring States period.

24 The stamped motifs on these wares are widely varied, ranging from designs that resemble fine woven textiles to large **checkers** arranged in fairly regular rows. A thin brownish luster that resembles a glaze often covers the surface of these utilitarian vessels.

25 **Architectural pottery**. Architectural pottery also had its place in the list of wares produced during this period. Among the building materials excavated at the remains of Warring States palaces and other buildings were water pipes and bricks, as well as ornamental roof tiles, **end tiles**, and waterspouts.

Text-Related Information

This text is adapted from *A Handbook of Chinese Ceramics* (1989), written by Suzanne G. Valenstein, published by The Metropolitan Museum of Art, New York and distributed by Harry N. Abrams, Inc., New York.

Selected Plates Related to the Text

Plate 6   Tripod. Earthenware with cord markings. Height: 16.2 cm. Late Shang dynasty.

Plate 7   Bowl. Earthenware with carved and incised decoration. Diameter: 22.5 cm. Western Zhou dynasty.

Plate 8　Duck-shaped vessel. Burnished earthenware. Eastern Zhou dynasty. Warring States era, end of 4th century B.C..

Plate 9　Jar. Earthenware with ribbing and cord markings. Height: 21.9 cm. Eastern Zhou dynasty.

Unit 2 Pottery Before the Qin and Han Dynasties

Plate 10   Jar. Earthenware with cord markings. Height: 25.7 cm. Eastern Zhou dynasty. Probably Warring States period, 475 – 221 B. C. .

Aids to Comprehension

Notes

(1) Shen-nung　神农氏,姜姓,中国上古人物,被世人尊称为"药祖""五谷先帝""神农大帝""地皇"等。华夏太古三皇之一,传说中的农业和医药的发明者。神农发现,经过火烧之后的土质容器会变得更加坚硬和耐用,于是陶器出现了,因此有神农发明陶器的传说。

(2) Triptolemus　特里普托勒摩斯,希腊神话人物。

(3) K'un-wu　昆吾,本名为樊,颛顼曾孙陆终长子,相传为陶器制造业的发明者。

(4) Yellow River　黄河,是位于中国北方地区的大河,属世界长河之一,中国第二长河。黄河是中华文明最主要的发源地,中国人称其为"母亲河"。

(5) Xia dynasty　夏朝(前2070—前1600年),是中国史书中记载的第一个奴隶制朝代。夏朝实际是由氏族为核心发展形成的国家,一般认为夏朝共传14代17后(夏统治者在位称"后",去世后称"帝"),建立者为大禹,定都阳城、斟鄩、安邑等地。河南嵩山一带和伊河、洛河流域为活动中心区。

(6) Erlitou culture　二里头文化,河南省洛阳市偃师二里头遗址一至四期所代表的一类考古学文化遗存,是介于中原龙山文化和二里岗文化的一种考古学文化,是华夏文明的重要组成部分。该考古文化主要集中分布于豫西、豫中,北至晋中,西至陕州区、丹江上游的商州地区,南至湖北北部,东至开封、兰考一带。二里头文化,既包括二里头遗址文化,又包括二里头遗址之外具有二里头遗址文化特征的上百处遗址所反映的文化面貌。

(7) Yanshi xian　偃师县,位于河南省洛阳市。

(8) Shanxi　山西省

(9) Shaanxi　陕西省

(10) Longshan cultures　龙山文化,泛指新石器时代晚期黄河中下游地区的一类文化遗存,属铜石并用的文化遗存。龙山文化因首次发现于山东省济南市历城县龙山镇(今属济南市章丘区)而得名。

(11) Zhengzhou phase　郑州商代时期。郑州在商朝被称为亳都,而亳都作为商朝都城的这一时期是商代早中期。

(12) Shang dynasty　商朝(前 1600 年—前 1046 年),是中国历史上的第二个朝代,也称殷商,是中国第一个有直接文字记载的王朝。商朝经历了三个大的阶段:第一阶段是"先商";第二阶段是"早商";第三阶段是"晚商"——前后相传 17 世 31 王,延续 500 余年。

(13) Yangtze River　长江,中国第一大河,也是中华民族的母亲河、中华民族的重要发祥地。

(14) Xiaotun　小屯村,即河南安阳小屯村遗址。

(15) Erlitou site　二里头遗址,全国重点文物保护单位,中华文明探源工程首批重点六大都邑之一。该遗址位于洛阳盆地东部的偃师区境内,遗址上最为丰富的文化遗存属二里头文化,其年代约为距今 3800—3500 年,相当于古代文献中的夏、商时期。该遗址南临古洛河,北依邙山,背靠黄河,范围包括二里头、圪垱头和四角楼等三个自然村,面积不少于 3 平方千米。作为全国重点文物保护单位,二里头遗址对研究华夏文明的渊源、国家的兴起、城市的起源、王都建设、王宫定制等重大问题具有重要的参考价值,是学术界公认的中国最引人瞩目的古文化遗址之一。

(16) Zhengzhou site　郑州商代遗址,位于河南省郑州市,东起凤凰台以东,西到西河口以西,北到花园路北段,南到二里岗及陇海铁路,是一座早于商代晚期都城——安阳殷墟的商代遗址群。

(17) Tunxi　屯溪区,位于安徽省境内。

(18) the Spring and Autumn era　春秋时期

(19) Ju　莒县,位于山东省。

(20) Junan xian　莒南县,位于山东省。

(21) the Warring States era　战国时期,是中国历史上继春秋时期之后的大变革时期,为列国诸侯争斗激烈的时代。

Ceramic Terminology

(1) remains　遗存;遗迹

（2）carbon-14 tests 碳十四测年法，通过测量含碳物质中放射性碳—14 的含量来确定其年代。这种方法在考古学中非常关键，因为它提供了一个相对精确的时间框架来研究不同地区的新石器文化和其他考古发现。总体而言，碳十四检测在考古学中是一种非常重要的工具。尽管它存在一些局限性和不准确性，但在适当的应用条件下，它仍然能够提供相对准确的年代信息。

（3）materials 出土物

（4）excavate: When archaeologists or other people excavate a piece of land, they remove earth carefully from it and look for things such as pots, bones, or buildings that are buried there, in order to discover information about the past. 挖掘（古物）

（5）black-slipped pottery 覆盖黑色化妆土的陶器。化妆土的作用主要是保护器物坯体，其颜色主要包括白色和黑色，白色的更普遍，黑色的较少使用。

（6）impress: to mark by or as if by pressure or stamping 压印，一种陶瓷纹饰手法，即将图案刻在模子中，然后将图案压印在陶瓷器物上；或者用拍子将图案拍印在陶瓷器物表面。

（7）hard earthenwares 硬陶。陶器按胎质可分为软陶和硬陶。硬陶就是烧造温度超过 1000 ℃的陶器，烧造温度在 1000 ℃以下的陶器称为软陶。

（8）"Cord-marked" pottery: A term used to refer to the relatively crude, utilitarian earthenwares used in ancient China. They may be unembellished or decorated in a simple manner; the surfaces frequently show deep, cordlike markings. 绳纹陶器。古代中国人使用的一种粗糙的日用陶器，这种陶器一般光素无纹或者装饰简单；器物表面通常呈现深深的压印绳纹。

（9）mold 模子。用刻有纹样的模子制坯，坯体上印有纹饰。

（10）beater-and-pad technique 拍打印纹技法，将刻有阴纹的模具拍印到陶器上，使陶器表面呈现出突起的阳纹；或者用刻有纹样图案的印戳或模子在尚未干透的坯体上拍印出花纹。

（11）finished 修坯工整的

（12）*li* tripod 鬲，是一种古陶器，用于蒸煮食物，其最基本的特征是三个肥大的足形似布袋。

(13) lacquered pottery　漆陶,指器表以漆料为颜料绘制图案的陶器。

(14) architectural pottery　建筑用陶器,简称建陶。

(15) unglazed pottery　未上釉的陶器,指器物表面没有覆盖釉层的陶器。

(16) high-footed bowl　高足碗

(17) incise: If an object is incised with a design, the design is carefully cut into the surface of the object with a sharp instrument. 划;雕;刻;划花

(18) glazed pottery　上釉陶器,指器物表面覆盖釉层的陶器。

(19) high-fired glaze　高温釉

(20) stoneware: vitrified, high-fired pottery made of clay to which a proportion of other materials may be added to achieve good working and firing properties. Stoneware is fired in excess of about 1200 ℃. It is dense, hard, resonant when struck, and impervious to liquid; it may be light or dark in color, but it is not translucent. 炻器,黏土搭配适量提高制作和烧制效果的其他材料,经过1200 ℃以上高温烧制而成的陶瓷。其胎体致密、坚固,不透水,敲击时发出响声,器表颜色或深或浅,不透明。

(21) stamp: If you stamp a mark or word on an object, you press the mark or word onto the object using a stamp or other device. 盖(章);印(某标记);打上(某字)。文中指压印,戳印纹饰,是陶瓷装饰手法,即将刻划好的图案印到陶瓷器物坯体的装饰手法。

(22) polish: to make smooth and glossy usually by friction. 磨光;精修

(23) mortuary pottery　陪葬陶器;明器

(24) painted unglazed pottery　未上釉彩绘陶器

(25) slip: a creamy mixture of clay, water and typically a pigment of some kind, used especially for decorating earthenware. 化妆土;泥浆

(26) burnished pottery: pottery made shiny or lustrous by rubbing　抛光陶器

(27) texture: The texture of something, especially food or soil, is its structure, whether it is light with lots of holes, or very heavy and solid. 纹理;肌理

(28) checker　方格图案;方格纹

(29) end tile　檐梢瓦

Exercises

Reading Comprehension Questions

Please answer the following questions according to the text.
(1) What are the features of geometric-impressed pottery in the Shang dynasty?
(2) What are the features of unglazed pottery in the Western Zhou period?
(3) What are the features of glazed pottery of the Spring and Autumn era?
(4) What are the features of burnished pottery of the Warring States era?
(5) What's the development of cord-marked pottery from the Shang dynasty to the Warring States Era?

Translation

A. Please translate the following paragraphs into Chinese, paying special attention to ceramic terminology, context and the social background involved.

(1) Western Zhou pottery provides a logical place to digress for a moment and discuss an important fact about the development of all Chinese pottery. Unlike automobiles from a Detroit assembly line, where this year's model is distinct from last year's or next year's car, Chinese ceramics exhibit a gradual, albeit constant, evolution in both style and physical characteristics, with relatively few major innovations interrupting this slow and measured pace. Throughout Chinese history, while the temper of the time certainly made itself felt in all creative efforts, changes in taste generally occurred by degrees. As might be expected from the empirical methods of manufacture that were used, the material qualities of fabricated objects altered progressively as well. It is therefore axiomatic that Chinese pottery—which on the whole developed in barely perceptible successive stages—usually cannot be pigeonholed in precise or narrow dates, either by Chinese or western calendars.

(2) Burnished pottery. The shapes of burnished gray wares with blackish surface were often copied from those made of other materials. Designs, which are slightly glossy and a little darker than the matte ground, were derived from those on contemporary bronzes and lacquers. While this technique of burnishing decorative figures on pottery originated in the Spring and Autumn era, it became especially popular during the Warring States period, when some extraordinary examples were produced. Burnished designs could be combined with patterns that were incised into the body before it was completely dry; and again, late Zhou gray wares might just be enriched with bronze-derived incised decoration.

B. Please translate the following paragraph into English.

人们最初制作陶器是不熟练的,只是用手捏成一些简单实用的器物。而制作一些大型的或小口大腹的较复杂的陶器器皿,必须在捏塑小型简单的器皿取得一定经验的基础上才能做到。手捏成形的方法,即泥条盘筑法,是将泥料捏成泥条,然后一只手拿着泥条螺旋式盘绕而上,做成陶坯的雏形。另一种制作方法是轮制方法,用这种方法做出的器物外形圆正规整,器物表面有一系列线条清晰的圆形纹路。还有一种方法叫范制法。陶范做出的陶器标准一致。

Unit 3   Three-color Glazed Wares of the Tang Dynasty

Tang Sancai is a renowned **low-fired** glazed pottery produced in China during the **Tang dynasty** (618 – 907 A. D.). The term *"sancai"* refers to the three colors used in the glazes, namely green, yellow and white. However, the colors of the glazes used to decorate the wares of the Tang dynasty were not limited in number to these three, as blue and brown examples were also made. During the Tang period, a large number of three-color glazed wares were exported to and imported from Central Asia and the Near East. These multi-colored glazed ceramics were viewed as especially prestigious in the Central Asia and Western Asia, increasingly becoming a popular style in Islamic ceramic. The Tang Sancai wares mainly include figurines for afterlife and wares for daily life. Although the manufacturing kilns and Tang Sancai wares are quite dispersed in the Tang empire, they are concentrated in the tombs and residential areas in the domain of **Chang'an**, the primary capital city at the modern city of Xi'an in Shaanxi province, and Luoyang, the secondary one at the synonymous city in Henan province.

Text
Three-color Glazed Wares of the Tang Dynasty
Lili Fang

Firing Techniques

1 The large-scale production of porcelain during the Tang dynasty to a large extent, pushed pottery out of people's daily lives, but pottery still firmly dominated burial object production and even saw new development, namely the invention of tricolor pottery. Tang tricolor pottery differed from ordinary low-temperature glazed pottery. Its body was made of white clay (**kaolinite**) and it had low-temperature glaze containing lead. The glaze used multiple metals including iron, copper, manganese and cobalt as **colorants**. Different metals had different coloring effects. For instance, after firing copper oxide was green, iron oxide was yellowish-brown, and cobalt oxide was blue. Lead was used as a **glaze solvent** to take advantage of its fluidity in the firing process to bring out a rich variety of colors including yellow, ochre, jade green, dark green, sky blue, brownish red, and eggplant purple. The gorgeous mixture of bright colors fills people's imagination with the splendor of the three-color glazed wares of Tang dynasty. The vessels were fired twice. First plain greenware was fired at about 1,100 ℃, then glaze was applied, and finally it was fired at a lower temperature of about 900 ℃. The technology was developed based on the **lead-glazed** pottery of the **Han dynasty**. The so-called "tricolor", a literal translation of its Chinese name, actually meant colorful or multicolor. In fact, the colors were in no way limited to three. Green, yellow, ochre, brown, red, white, blue, black and many other colors could all be seen, sometimes only one for one ware but in most cases several in a mixture.

2 The origin of the tricolor pottery of the Tang dynasty can be found in the lead-glazed pottery of the Han dynasty. Back in the Han dynasty Chinese potters succeeded in making monochrome glazed pottery with a glossy and brightly colored appearance. Later in the **Southern & Northern dynasties**, lead-glazed pottery in the north witnessed new developments, taking a step further from the monochrome

glaze of the Han dynasty to the firing of ware with a white background and green coloring, ware with a yellow background and green coloring, and triple-color ware of yellow, brown and green. Thus, it is safe to say that the emergence of the Tang tricolor ware was an inevitable result of the development of the glazed pottery of the **Han and Wei dynasties**, yet the magnificence, splendor and unusual effervescence of Tang dynasty tricolor pottery was far from something that the Han dynasty glazed pottery or any other painted pottery could compare with.

Functions and Decorative Techniques

3 Because of their high lead content, Tang tricolor wares were seldom used in daily life and were mostly burial objects. They can be roughly divided into three categories: vessels, figurines and models. There were over 10 varieties of vessel, including bottles, pots, jars, bowls, cups, plates, *yu*, candlesticks, and pillows. Each variety had many styles. For bottles alone there were double-dragon-handled bottles, **double-lugged flat bottles**, flowery mouthed bottles, washer-style-mouthed bottles, and thin-necked melon-belly bottles. **Figurines** were of both human and animal, the former including ladies, female and male servants, horse-leading men, civil officials, warriors, foreigners, and **guardians of Buddhism** and the latter including horses, donkeys, camels, pigs, cattle, sheep, dogs, chickens, and ducks. The third category was models of various architectures and household utilities, covering almost everything that the dead may have enjoyed when alive, with the most common being buildings, towers, gardens with rockery and waterside pavilions, and all kinds of houses, warehouses, toilets, carriages, and cabinets. Tricolor figurines came into existence later than tricolor vessels and appeared in large numbers in tombs only after the reign of **Wu Zetian**. Tang tricolor pottery, overcoming the limitation of previous monochrome glazes, used a variety of glaze colors for decoration and achieved an extraordinarily exquisite artistic effect. Although the surface decoration techniques of Tang tricolor ware were traditional, like printing, decals, **engraving**, and **sculpting**, it had its own unique prominence. Potters used a brush that had been dipped in glaze to draw reticulate patterns, speckles and colorful ribbons on the vessels, or, imitating the batik process of Tang

dynasty silk, put **wax** on certain areas of the vessel surface and then applied glaze so that the waxed areas were not colored by glaze and retained the original white body, which set off and added radiance to the colorful and brilliant patterns. Another common decorative technique of tricolor pottery was the application of bright colorful glazes on sculptured or **imprinted** patterns to create an effect similar to that of **relief**. The development of tricolor decorative art in the Tang dynasty was influenced not only by ceramic art itself, but also, directly or indirectly, by contemporaneous sister art forms such as painting, gold and silver ware, sculpture, lacquerware, and **shell embedding** technology.

4 The clay used by Tang tricolor pottery had excellent plasticity and therefore allowed rich molding methods, among which the most commonly used were **clay slice bonding**, the potter's wheel and fashioning with models. Objects fashioned by clay slice bonding were mostly square or rectangular, such as pillows and couches, while bottles, jars, plates, bowls and dishes were shaped with the wheel.

Tricolor Pottery Statues

5 Burying pottery statues with the dead had long been popular before the Tang dynasty, notably the **terracotta warriors** and horses of the **Qin dynasty** and the various human and animal figurines of the Han dynasty. Such a funerary custom continued into the Tang dynasty and as a result Tang tricolor pottery as burial objects experienced rapid development, especially pottery statues, which, with their rich varieties and wide range, can serve as a key for us to learn about and understand the culture of the Tang dynasty. Mostly unearthed from the tombs of high officials and dignitaries, the figurines represented the image of noblewomen in the Tang dynasty, with some the symbols of the tomb owner. Under the influence of the imperial court's pursuit of luxury, the clothes of noble women in the Tang dynasty were gorgeously colored and meticulously collocated. The dresses of the tricolor pottery figurines unearthed in Tang tombs in Shaanxi and Henan are breath-taking. They are accurate reproductions of women's clothes at the time. The prosperity of the economy and culture in the Tang dynasty brought about women's liberation. A large amount of materials unearthed from the tombs of the Tang dynasty showed that women at that

time could follow their own interests in dress, and were also allowed to go on excursions or go hunting on horseback, play polo, and participate in other sports activities like men. The large number of horse-riding women figurines unearthed from tombs of the Tang dynasty were a true portrayal of the social fashion.

6 Horse statues make up a large proportion of the cultural relics unearthed from the Tang dynasty, and among them tricolor pottery horses are an especially prominent part. Tang tricolor horses had a strong style unique to the time with small heads, long necks, and plump bodies. They are of two sizes, large between 69 and 81 cm and small about 30 cm. The horses came in multiple colors, including blue, yellow, white, black and green, with neatly trimmed bristles and saddles. The saddles had green mud barriers, and some were decorated with green tassels. Some horses had decorative accessories on their bodies and heads, such as apricot leaf shaped patterns on the headstall and the same pattern or blossoms on the leather straps. Statues of horse dancing, horse-leading figurines and horse training figurines were also common.

7 In addition, figurines of civil officials, military officers, guardians of Buddhism, and tomb-guarding beasts also frequently appeared in the tombs of the Tang dynasty nobility, and occupied an important position in the funerary culture of the Tang dynasty. Figurines of civil officials had kind expressions, neat clothes, and standard postures, in general a sage-like, gentle, respectful and modest demeanor. Figurines of military officers, on the contrary, were daunting in both facial expression and clothing. They were symbols of force and majesty, the nemesis of demons and ghosts, and defenders of the underworld. The guardians of Buddhism, according to legend, could not only protect Buddhist altars from demons, but also protect tombs from evil. With the rising popularity of Buddhism, production of such figurines became prevalent in the Tang dynasty and the figurines were very popular in large and medium-sized tombs of the Tang dynasty. Figurines of tomb-guarding beasts, mostly squatting, with human or animal faces, large curved horns and big ears on their heads, and wings and flames on their shoulders and backs. They were ferocious and believed to drive away vile forces, repel epidemic diseases, and keep the soul of the tomb owner at peace.

8   In addition to human and animal statues, there was another type of burial object, that is, daily life utensils, which guaranteed that the dead would have all the vessels they would require in the netherworld. For instance, unearthed from **Crown Prince Zhanghuai**'s Tomb there were tricolor bowls, cups, **thin-necked bottles**, ink stones, ochre glazed bowls, *dou*, plates, incense burners and **tripod stoves**. The vessels, just like the statues described above, displayed exquisite craftsmanship and the distinct characteristics of their time. There were also innovative forms such as phoenix headed pots, double-dragon *zun*, tower-style jars, duck-shaped cups, dragon-headed cups, elephant-headed cups, mandarin duck cups, and double-fish pots, which were a real demonstration of the innovative artistic quality of the Tang people and the broad and profound style of Tang culture and art.

Text Related Information

This text is adapted from chapter seven of the book *The History of Chinese Ceramics* (2013), written by Lili Fang, published by Foreign Language Teaching and Research Press, Beijing.

Selected Plates Related to the Text

Plate 11   Tricolor plate with impressed design. Diameter: 28.9 cm. Tang dynasty.

Plate 12 Tricolor pheonix-head pot. Height: 33 cm, top diameter: 5.7 cm, base diameter: 10.4 cm. Tang dynasty.

Plate 13 Tomb guardian and earth spirit, earthenware with "three-color" glazes and pigments. Height of tomb guardian: 88.9 cm, height of earth spirit: 78.1 cm. Tang dynasty.

Unit 3 Three-color Glazed Wares of the Tang Dynasty

Plate 14 Figure of lady, earthenware with white slip and remains of pigments. Height: 40.6 cm. Tang dynasty.

Aids to Comprehension

Notes

（1）Tang dynasty　唐朝（618—907年），是中国历史上继隋朝之后的大一统中原王朝，共历21帝，享国289年。唐朝统治阶级以开放的心态与博大的胸襟接纳外来文化，日本、南诏、新罗、渤海等国均遣使学习唐的制度、文化，各民族呈现大融合局面。唐朝国力强盛，经济繁荣，对外交往活跃，在制度上也有重要建树，是当时世界上最强盛的国家之一。唐以后海外多称中国人为"唐人"。通过与各国的交流，唐朝的经济、社会、文化、艺术呈现出多元化、开放性等特点，在诗歌、书法、绘画、音乐等方面均取得了重要的成果。

（2）Chang'an　长安，今陕西省西安市。

（3）Han dynasty　汉朝（前206—220年），是继秦朝之后的大一统王朝，分为西汉（前206—25年）、东汉（25—220年）两个时期。汉朝在思想、史学、文学、艺术、科技等方面卓有建树，儒学独尊、佛道并列，汉赋、乐府诗兴盛，《史记》《汉书》开创纪传体史书先例，造纸术的发明与改进更推动了人类文明的发展。

（4）the Southern & Northern dynasties　南北朝（420—589年），是南朝（420—589年）和北朝（386—581年）的统称，指中国历史上从420年东晋灭亡、南朝建立到589年隋朝统一全国形成的南北对峙局面。南方四朝社会相对稳定，江南开发规模空前，南方经济得到长足发展。南北朝时期，社会出现民族大融合的趋势，统一的多民族国家得以重生，为此后隋唐时期的繁荣强盛奠定基础。

（5）the Han and Wei dynasties　汉朝和魏朝。汉朝，见前文注释。魏国（220—265年）又称曹魏，是中国汉朝末期三国之中最强大的一个政权，由魏文帝曹丕建立，定都洛阳。

（6）guardians of Buddhism　天王。在佛教中，"四大天王"是指四位护法神，分别是持国天王、增长天王、广目天王和多闻天王。

(7) Wu Zetian　武则天(624—705年),名武曌,并州文水县(现山西省文水县东)人,中国历史上唯一的女皇帝。

(8) shell embedding　螺钿,是贝壳类的镶嵌装饰工艺,通常又称螺填、钿螺等。螺钿是我国独有的传统艺术瑰宝,通常被应用于漆器、家具、乐器、插屏以及木雕一类的工艺品上。其精髓在于螺的加工,即根据需要将螺壳打磨加工成人物、花鸟或文字等薄片,镶嵌到工艺品的表面,可以形成强烈的视觉冲击。

(9) terracotta warriors　兵马俑即制成兵马(战车、战马、士兵)形状的殉葬品。1961年3月4日,秦始皇陵被国务院公布为第一批全国重点文物保护单位。1974年1月,兵马俑被发现。1987年,秦始皇陵及兵马俑坑被联合国教科文组织批准列入世界遗产名录,被誉为"世界第八大奇迹"。

(10) Qin dynasty　秦朝(前221—前207年),是中国历史上第一个统一的封建王朝。秦朝结束了春秋战国以来诸侯分裂割据的局面,成为中国历史上第一个中央集权制国家,对中国历史产生了深远影响。秦朝的疆域,东到东海,西到陇西,北到长城一带,南到南海,大大超过了前代。

(11) Zhanghuai's Tomb　章怀太子墓,位于陕西省咸阳市乾县,是唐高宗李治和武则天的次子——章怀太子李贤和太子妃房氏的合葬墓,也是高宗乾陵的陪葬墓之一。章怀太子墓园长180米,宽143米。封土为覆斗形,底边正方,边长43米,高18米。地宫全长71米,宽3.3米,深7米。该遗址出土壁画、彩绘陶俑、三彩器、陶瓷器和石刻等文物600余件,对研究唐代历史和文化艺术提供了极为重要的实物资料。1961年3月4日,包含章怀太子墓在内的乾陵被国务院公布为第一批全国重点文物保护单位。

Ceramic Terminology

(1) Tang sancai ware(s): tricolor ware(s), three-color glazed ware(s)　唐三彩,中国古代陶瓷烧制工艺珍品,全名唐代三彩釉陶器,是盛行于唐代的一种低温釉陶器。釉彩有黄、绿、白、褐、蓝、黑等色彩,以黄、绿、白三色为主,所以人们习惯称之为"唐三彩"。因洛阳唐三彩最早出土且出土数量最多,亦有"洛阳唐三彩"之称。

(2) low-fired 低温烧制。低温陶瓷的烧制温度通常在 800 ℃ 以下,而高温陶瓷的烧制温度通常在 1300 ℃ 以上。低温陶瓷和高温陶瓷在烧制温度、颜色、透明度、硬度等方面都存在差异,适用于不同的领域。

(3) kaolinite 高岭石,一种陶土。

(4) colorant 着色剂。文中指陶器彩绘所用的矿物着色剂,主要包括铁、铜、锰、钴等。

(5) glaze solvent 釉料助熔剂。助熔配料是指一种促进熔化或玻璃化的物质,能降低物质或混合物的熔点。用于釉料或瓷釉中的助熔剂有长石、铅白、铅丹和硼砂。

(6) lead-glazed 施铅釉的。铅釉是釉料中的一种,为低温釉料。

(7) yu 盂,一种器皿,用于盛饮食或其他液体的圆口器皿。

(8) double-lugged flat bottle 双系扁瓶

(9) figurine 小雕像。文中指陶瓷雕像,包括人物塑像和动物塑像。

(10) engraving 刻花。文中指陶器表面的装饰手法之一,即在器物表面雕刻花纹。

(11) sculpting 塑像法。这里指用捏制或模制等方法制作陶器塑像的方法。

(12) wax 蜡。这里指在器物胎体表面某些区域覆盖蜡,其他地方施釉。覆盖蜡的地方保留了原来胎体的颜色,与施釉区域的釉色形成对比,产生装饰效果。

(13) imprinted 印花,陶瓷装饰方法之一,将图案压印到器物表面,产生浮雕效果。

(14) relief 浮雕

(15) clay slice bonding 泥片围接法,陶瓷塑形手法之一,即将泥块拍打成需要大小和厚度的泥片,然后再根据器型拼接在一起,最后得到所需要的形状。常见的陶瓷成型方法还包括:手工捏塑法,即手工将泥巴捏成不同的造型,这最原始的成型方法;泥条盘筑法,即将陶泥搓成条状,从底部开始,一层层堆叠成器物的形状,再用泥浆黏合成一个整体的器物,最后磨平盘筑时的痕迹,新石器时代的很多陶器就是用这种方法制作的;轮制成型法(拉坯),是将泥料放在特制的轮盘上,利用轮盘转动所产生的离心力使坯体成型的技法,这种方法主要用来做各种圆形的器物;印模法,即将泥料压入特制的模具中使之成型,这种方法主要用来制作小动物、一些大型器物的零部件或者纹饰以及陶瓷人俑等;灌浆法,即把泥制成泥浆,然后灌入事先设计好的石膏模型之中,石膏能够将泥浆中的悬浮物吸附于内壁,最终成型,这是目前大多数日用瓷的成型方法。

(16) thin-necked bottle　细颈瓶

(17) dou　陶豆,形状像高脚盘,是中国先秦时期的食器和礼器,流行于春秋战国时期,开始时用于盛放黍、稷等谷物,后用于盛放腌菜、肉酱等。

(18) tripod stove　三足炉

(19) zun　尊,是中国商周时代的一种大中型盛酒器。其形制为圈足,圆腹或方腹,长颈,敞口,口径较大。尊盛行于商代至西周时期,春秋后期已经很少见。较著名的有四羊方尊。

Exercises

Reading Comprehension Questions

Please answer the following questions according to the text.

(1) How did Tang tricolor pottery differ from ordinary low-temperature glazed pottery?

(2) What are the three categories of Tang tricolor pottery?

(3) What are the surface decoration techniques of Tang tricolor ware?

(4) How did tricolor pottery statues reflect the social life of the Tang dynasty?

(5) How did tricolor pottery statues reflect the funerary culture of the Tang dynasty?

Translation

A. Please translate the following paragraphs into Chinese, paying special attention to ceramic terminology, context and the social background involved.

(1) Because of their high lead content, Tang tricolor wares were seldom used in daily life and were mostly burial objects. They can be roughly divided into three categories: vessels, figurines and models. There were over 10 varieties of vessel, including bottles, pots, jars, bowls, cups, plates, *yu*, candlesticks, and pillows. Each variety had many styles. For bottles alone there were double-dragon-handled bottles, double-lugged flat bottles, flowery mouthed bottles, washer-style-mouthed bottles, and thin-necked melon-belly bottles. Figurines were of both human and animal, the former including ladies, female and male servants, horse-leading men, civil officials, warriors, foreigners, and guardians of Buddhism and the latter including horses, donkeys, camels, pigs, cattle, sheep, dogs, chickens, and ducks. The third category was models of various architectures and household utilities, covering almost everything that the dead may have enjoyed when alive, with the most common being buildings, towers,

gardens with rockery and waterside pavilions, and all kinds of houses, warehouses, toilets, carriages, and cabinets.

(2) Tang Sancai is a renowned low-fired glazed pottery produced in China during the Tang dynasty (618 – 907 A.D.). The term "*sancai*" refers to the three colors used in the glazes, namely green, yellow and white. However, the colors of the glazes used to decorate the wares of the Tang dynasty were not limited in number to these three, as blue and brown examples were also made. During the Tang period, a large number of three-color glazed wares were exported to and imported from Central Asia and the Near East. These multi-colored glazed ceramics were viewed as especially prestigious in the Central Asia and Western Asia, increasingly becoming a popular style in Islamic ceramic. The Tang Sancai wares mainly include figurines for afterlife and wares for daily life.

B. Please translate the following paragraph into English.

三彩陶器是唐代陶瓷中具有特殊作用和风格的一朵奇葩。在20世纪初期的一次考古发现中,人们在河南洛阳附近的一座古墓中,发现了大量的"唐三彩"明器,其独具风采的艺术形象再次驰名中外。"唐三彩"陶器是在汉代铅釉陶的基础上进一步发展起来的,用白色黏土做胎,施以含铅的低温釉,釉中使用铁、铜、锰、钴等多种金属作呈色剂,在750 ℃—800 ℃的低温下焙烧而成。所谓"三彩"即多彩之意。

Unit 4 Song Dynasty Kilns in Northern China

The Song dynasty, established in 960 A. D. , ushered in a prolonged era of prosperity marked by the cultivation of arts. Despite border contractions and Tartar threats, the dynasty thrived, relocating its capital to Lin'an in 1138 after the fall of the northern capital. Lin'an, now known as Hangzhou, praised by Marco Polo as the world's finest city in 1280, exemplified Song civilization's refinement and luxury with its canals, bridges, guilds, markets, and hospitable citizens. This era fostered the development of literature, art, and connoisseurship, with treatises, encyclopedias, and illustrated catalogues being published. The Song dynasty excelled in ceramics, with imperial patronage boosting the art form. Factories like Dingzhou, Ruzhou, and the renowned Jingdezhen produced exquisite wares well-known for their glazes, often decorated through moulding, stamping, and relief carving before firing. The glazes, rich in color and texture, defined Song ceramics' appeal.

The Song culture was introverted due to external threats, fostering a peaceful, introspective atmosphere reflected in its ceramics. Court ceramics exemplified quiet elegance, while more common wares catered to popular tastes with bolder colors and decorations. The Song dynasty's artistic achievements, particularly in ceramics, continue to inspire and be revered in later Chinese art.

Text

Ting Kilns

R. L. Hobson

1 As already hinted in the passage above, Ting ware suffered a temporary eclipse at Court owing to some defects in the glaze; but it was not long in recovering its reputation, for the *Ko ku yao lun* states that it was at its best in the Zhenghe and Xuanhe periods, which extended from 1111 A. D. to 1125 A. D. and we learn that the Ting Chou potters accompanied the Court in its flight across the Yangtze in 1127. The manufacture seems to have been re-established after this event in the neighbourhood of Ching-te-chen, and the Nan Ting or Southern Ting ware is said to have so closely resembled the original that to distinguish the two in after years was regarded as a supreme test of connoisseur-ship.

2 Ting ware has a white body of fine grain and compact texture, varying from a slightly translucent porcelain to opaque porcellanous stoneware. Though not so completely vitrified as the more modern porcelains, and lacking their flint-like fracture, it was nevertheless capable of transmitting light in the thinner and finer specimens, wares which fulfills the European definition of porcelain. The glaze is of ivory tint, sometimes forming on the outsides of bowls or dishes in brownish gummy **tears**, which were regarded by Chinese collectors as a sign of genuineness. The finer and whiter varieties are known as *Pai Ting* (white Ting) and *Fen Ding* (flour Ting), as distinct from the coarser kind, whose opaque, earthy body and glaze of yellowish tone, usually crackled and stained, earned it the name of Tu Ting or Earthen Ting. In the best period the pure white undecorated Ting ware, with rich unctuous glaze, compared to "congealed fat" or "mutton fat," was most esteemed, though ornament was freely used, especially on the Southern Ting. Designs carved in low relief or etched with a point were considered best, the molded and stamped ornament being rightly regarded as inferior.

3 Favourite carved designs with the Ting potters seem to have been the mu-tan peony, the lily, and flying phoenixes. They are, at any rate, usually singled out for

mention by Chinese writers. Garlic and rushes are also incidentally mentioned as motives, and a few examples of a beautiful design of ducks on water are known in western collections. The molded ornament which is generally more carved designs borrowed from ancient bronzes must have been highly prized. Of the three kinds of ornament usually associated by Chinese writers with the Ting ware, the **hua hua** (**carved decoration**) and the **yin hua (stamped or moulded decoration)** have already been mentioned. The meaning of the third, **hsiu hua**, is not so clear, as the phrase can bear two interpretations, viz. painted ornament or embroidered ornament. In the latter sense it would suggest a rich decoration like that of brocade without indicating the method by which it was applied. But in the former it was the usual Chinese expression for painted ornament, and it is difficult to imagine that it was intended to indicate anything else in the present context. On the other hand, no examples of painted Ting ware are known to exist either in actual fact or in Chinese descriptions. This anomaly, however, may perhaps be explained in one of two ways. A creamy white ware of Tu Ting type, boldly painted with brown or black designs, is known to have been made at the not far distant factories of **Tz'u Chou** in the Song dynasty, and it is possible that either the painted Ting ware has been grouped with the Tz'u Chou ware in modern collections, or that Chinese writers mistook the Tz'u Chou ware for painted Ting ware and added this third category to the Ting wares by mistake. In any case they regarded the painted ware as an inferior article.

4 The high estimation in which fine specimens of white Ting ware have always been held by Chinese connoisseurs is well illustrated by a passage in the *Yun shih chai pi tan*. It tells how Mr. Sun of the Wu-i river estate treasured in his mountain retreat Ting Yao incense-burners, and among them one exquisite specimen of the Song period. It was a round vessel with ear handles and three feet, and the inscription "李西涯" (Li Hsi Ya) was engraved in seal characters on the stand. During the Japanese raids in the Jiaqing period this vessel passed into the hands of one Chin Shang-pao, who sold it to T'ang, the President of Sacrifices(Tai ch'ang) of P'i-ling. T'ang, whose residence bore the romantic but chilly name of *Ning-an* (Frozen Hut), is the celebrated collector. "T'ang had many wonderful porcelains," the story runs,

"when this vessel arrived, they all, without exception, made way for it. And so throughout the land when men discuss porcelains, they give the first place to T'ang's white incense vase. " "T'ang, they say, did not readily allow it to be seen. " And in this respect, if all accounts are true, T'ang was not unlike a good many Chinese collectors of the present day.

5 The Pai Ting and the T'u Ting, the fine and coarse white varieties, alone have been identified in western collections; but there are colored Ting porcelains which are known to us by literary references. An apocryphal red Ting ware (Hung Ting) is mentioned in two passages of ambiguous meaning which need not necessarily have implied a true red glaze. In any case it finds no place in the older works, such as the *Ko ku yao lun* and *Ch'ing pi tsang*, which only speak of purple or brown (**tzu**) Ting, and black Ting. "There is purple Ting, " says the *Ko ku yao lun*, "the color of which is purple; there is ink Ting, the color of which is black, like lacquer. The body in every case is white, and the value of these is higher than that of white Ting. "

Text Related Information

The text is adapted from chapter seven of the book *Chinese Potter and Porcelain* (1915), written by R. L. Hobson, published by Funk and Wagnalls Company, New York.

Ru Kilns

R. L. Hobson

6 The Ru Yao was the porcelain made during the Song dynasty at **Ju-chou**, in the province of Honan, the modern Ju-chou-fu. Though no authenticated example of Ju ware is known in Europe, it is impossible to ignore a factory whose productions were unanimously acclaimed by Chinese writers as the cream of the Song wares. Ju Chou lies in the very district which was celebrated in a previous reign for the Ch'ai pottery, and it is probable that the Ju factories continued the traditions of this mysterious

ware. Nothing, however, is known of them until they received the imperial command to supply a *ching* (blue or green) porcelain to take the place of the white Ting Chou porcelain which had fallen into temporary disfavour on account of certain blemishes. This event, which took place towards the end of the Northern Song period (960 – 1127 A. D.), implies that whatever had been their past history, the Ju Chou factories were at this period pre-eminent for the beauty of their ching porcelain. It would appear from the **Ching po tsa chih**, which was written in 1192, that the Ju Chou potters were set to work in the "forbidden precincts of the Palace," and that selected pieces only were offered for imperial use, the rejected specimens being offered for sale. Even at the end of the twelfth century it was believed to be very difficult to obtain examples of the ware.

7 From the various accounts on which ceramic experts have to depend for the conception of the ware, it is clear that the body was of a dark color. The glaze was thick and of a color variously described as "approaching **the blue of the sky after rain**" (i. e. like the Ch'ai ware), pale blue or green, and "**egg white**" which seems to imply a white ware with a faint greenish tinge. The author of the **Ching pi tsang**, a work of considerable repute published in 1595, gives a first-hand description of the ware: "Ju Yao I have seen. Its color is 'egg white' and its glaze is lustrous and thick like massed lard. In the glaze appear faint 'palm eye' markings like crabs' claws. Specimens with sesamum designs (lit. flowers), finely and minutely engraved on the bottom, are genuine. As compared with Kuan Yao in material and make, it is more rich and unctuous (**tzu run**)." Two mysterious peculiarities have been attributed to the Ju ware, viz. that powdered **cornaline** was mixed with the glaze, and that a row of nail heads was sometimes found under the base. The first has been taken as merely an imaginative explanation of the lustre of the glaze, but it is certain that some kind of pulverized quartz-like stone was used in the composition of later glazes, such as the "**ruby red**". The second, which has been seriously interpreted to mean that actual metal nails were found protruding from the glaze (a physical impossibility, as the metal would inevitably have melted in the kiln), is probably due to a misunderstanding of a difficult Chinese phrase, **cheng ting**, which may

mean "engraved with a point" or "cut nails." The former seems to satisfy the requirements of the case, though it would be possible to render the sentence, "having sesamum flowers on the bottom and fine small nails," referring to the little projections often found on the bottom of dishes which have been supported in the kiln on pointed rests or "spurs."

8 Probably the safest clue to the appearance of Ju ware is to be found in the important passage which was written by a Song writer named Xu Jing who described the Corean wares as in general appearance like the old **Pi-se ware** of Yueh Chou and the Ju Chou ware. The typical Corean wares of this time are not uncommon, and their glaze-ash soft grey green or greenish grey, with a more or less obvious tinge of blue—would satisfy the Chinese phrases, tan ch'ing and fen ch'ing, and in the bluer specimens might, by a stretch of poetic phrase, even be likened to the sky after rain. The "egg white", however, must have been a somewhat paler tint if the expression can be taken in any literal sense.

9 From the foregoing considerations, it may be concluded that the Ju porcelain was a beautiful ware of celadon type, varying in tint from a very pale green to a bluish green.

10 Though it is nowhere definitely stated how long the Ju Chou factories retained their supremacy, it is tolerably clear from Hsu Ching's reference in 1125, or very soon after, to the "modern porcelain of Ju Chou," that they came into prominence towards the end of the Northern Song period, perhaps in the last half of the eleventh century; and as we have no further information about them, we may perhaps infer that they sank into obscurity when the Song emperors were driven from the North of China by the invading Tartars in 1127. In any case, the Ju ware seems to have become as extinct as the Ch'ai by the end of the Ming dynasty. Hsiang Yuan-p'ien, late in the sixteenth century, states that "Ju Yao vessels are disappearing. The very few which exist are almost all dishes, cups, and the like, and many of these are damaged and imperfect." It is not to be supposed that Ju Chou had the monopoly of the particular kind of ch'ing ware in which its factories excelled.

11 A number of other and not distant potteries were engaged in a similar

manufacture, though with less conspicuous success. It is reported according to **Cho keng lu** that "it was made in the districts of T'ang, Teng, and Yao on the north of the (Yellow) River, though the productions of Ju Chou were the best."

Text Related Information

This text *Ru Kilns* is adapted from chapter five of the book *Chinese Potter and Porcelain* (1915), written by R. L. Hobson, published by Funk and Wagnalls Company, New York.

The Cizhou Kilns

Lili Fang

12 The Cizhou kilns in modern-day Cixian county, Hebei Province, were prominent folk kiln clusters during the Song dynasty. Named after the jurisdiction of Cizhou over the Zhanghe and Fuhe river basins where they were located, these kilns encompassed centers like Guantai, Linshui, and Pengcheng. Cizhou kiln products featured brush painting on white glaze in various hues, utilizing techniques like carving, **sgraffito**, and filling to innovate porcelain decoration. Tracing its roots to Tang dynasty kilns, Cizhou kilns maintained this diversity, incorporating under-glaze black and brown painting influenced by Changsha kilns. This led to the formation of a broader Cizhou kiln system with a shared style adopted by many northern folk kilns.

13 Cizhou kiln sites that have been discovered over years of investigation and excavation are mainly concentrated in two areas. One group of sites is located on both banks of the Zhanghe river, six kilometers west of **Guantai town**, and the other group has **Pengcheng town** as its center. Both groups had densely distributed sites with large quantities of porcelain shards and remains buried underground.

14 Cizhou kilns started producing celadon in the Northern dynasties and white porcelain in late Five dynasties. In Song dynasty, Cizhou kilns, with their unique

variety of decorative arts, produced practical life utensils with a strong folk interest that were deeply loved by middle & lower classes. The decorative content of Cizhou kiln ware contains very rich northern folk customs and is therefore sometimes referred to as a folk culture museum or "folk culture dictionary".

15 The body of Cizhou kiln ware was made of the local raw material "**rough blue earth**," inferior to the material for the white porcelain body of Ding and Xing kiln ware. In response to this "congenital deficiency," the craftsmen of the Cizhou kilns first applied a layer of **white engobe** on the body, and then used various decorative techniques, such as carving, engraving, sgraffito, printing, painting, and glazing, to decorate the products. Especially salient was the adoption of traditional Chinese ink painting techniques to draw black patterns on a white porcelain background with a writing brush. The technique, with the strong contrast between black and white, created a bright, vivid and expressive artistic effect. An extremely free, unrestrained and rough painting style was adopted to depict images of traditional folk customs and interests popular among the people, thus forming the simple, unrestrained, and bold decorative language and style of the Cizhou kilns, a unique decorative art that prepared technological and material conditions for the transition of China's porcelain decoration from glaze ornament to color painting and for the emergence and development of blue and white ware, wucai ware and doucai ware in China.

16 It was noted in *A History of Chinese Ceramics* that "the Guantai kiln was the most representative Cizhou kiln, with its product lines and varieties being the epitome of all kilns in the Cizhou kiln system," a comment that was proven tenable by archaeological excavations of the Guantai and Pengcheng kilns in 1987 and 1999 as well as many investigations of other ancient kiln sites.

17 Of all the Cizhou kilns, the Guantai kiln produced the richest types of porcelain. In addition to white glaze ware and black glaze ware, there were a further 12 types, namely, white glaze with carved patterns, white glaze with sgraffito, white glaze with green spots, white glaze with brown spots, black painting under white glaze, caramel painting under white glaze, caramel painting and carved patterns under white glaze, carved patterns against a pearl ground, painting under a green

glaze, white glaze with red-green painting, low-temperature lead glaze, and tricolor ware. The rich varieties made Guantai the most representative kiln site in the Cizhou kiln system.

18 Based on the 1987 excavation report on the Guantai kiln by the Department of Archaeology of Peking University, we could preliminarily divide the development of the Guantai kiln into three stages. The first stage, the founding stage of the Cizhou kilns, was in the early Northern Song dynasty. Varieties were mainly daily use vessels such as bowls, plates, incense burners, water pots, bottles, and pillows. The most common decoration was white glaze with carved patterns. The patterns were dominated by lotus flowers, plants with rolled leaves, and water ripples. In between the patterns, parallel lines were carved with comb-like tools. Products with carved patterns against a pearl ground were also very distinctive. In addition to white glaze ware, there were small numbers of black glaze ware, blue glaze ware and sesame glaze ware.

19 The second stage, the developing period of the Cizhou kilns, was in the middle and late Northern Song dynasty. In this period varieties increased and incorporated richer changes. There emerged bowls and other utensils imitating Ding kiln products. Previous decorative techniques were inherited, white glaze with sgraffito was developed and black glaze with sgraffito appeared. Sgraffito refers to cutting out the ground surrounding the patterns so that the patterns appeared in relief. Stamens were then carved on the flowers and veins on the leaves. Cutting out the ground revealed the yellowish-brown body color, which highlighted the white or black decorative patterns. By this time a pearl ground with carved patterns became a common decorative method of the Cizhou kilns and the pearl ground of the Guantai kiln was the most beautiful with an **orange-red hue**. In addition, a very small amount of under-glaze black painting ware appeared, whose decorative themes included Chinese characters, human figures, and animals.

20 The Cizhou kilns had a significant impact, forming a system spanning Hebei, Henan, and Shanxi. The Dengfeng kiln, a Song dynasty porcelain kiln, produced pearl ground porcelain with carved patterns, inspired by Cizhou techniques.

Duandian kiln in Henan also crafted similar wares, though with less refined workmanship and less color filling due to larger sizes. Duandian notably produced more inscribed pillows and marked products with dates and workshop names.

Text Related Information

This text *The Cizhou Kilns* is adapted from chapter eight of the book *The History of Chinese Ceramics* (2013), written by Lili Fang, published by Foreign Language Teaching and Research Press, Beijing.

Chun Yao

Stephen W. Bushell   Lili Fang

21   Chun Yao is the name given to the porcelain fabricated at Chun-chou from the early part of the Song dynasty, which began in the year 960 A. D.. This corresponds to the modern district of Yun-chou in the province of Honan. It was not ranked high among the potteries of the period, because the material was not so finely levigated, and because the forms were generally original, instead of being copied from classical designs. The glazes were, however, remarkable for their brilliancy and for their varieties of color, including as they did the flambe or transmutation glazes, composed of flashing reds, passing through every intermediate shade of purple to pale blue. This was not much appreciated at the time, being described as a failure in the firing of one of the pure monochromes, but its reproduction in the hands of more recent potters is universally regarded as one of the chief triumphs of Chinese ceramic art.

22   The author of the ***Po wu yao lan***, one of the best of the antiquarian works published near the end of the Ming dynasty, in sixteen books, says in the fifth book, which is the one devoted to ceramics: "Chun-chou porcelain includes pieces of **vermilion red**, of **bright onion-green**, vulgarly called **parrot-green**, and of **aubergine purple**. When these three colors, the first red as mineral rouge, the

second green as onion sprouts or kingfisher feathers, and the third purple dark as ink-black, are pure and without the least change of color. They comprise the highest class. Underneath the piece one or two numerals are often inscribed as marks. The colors of pig's liver, of flaming red, and of blues and greens mingled in blotches like a child's tear-stained face, are due only to insufficient firing of the above three colors; they are not distinct varieties of glaze. Such vulgar names as 'nasal mucus' and 'pig's liver' only provoke ridicule. The flowerpots and saucers of this porcelain are of great beauty, but the other things, like the barrel-shaped seats, the censers and round pots for incense, the square vases and jars with covers, all these have the paste composed of yellow sand, so that they are of coarser fabric. The new pieces made in the present day are all fabricated out of Yi-hsing clay, so that, although the glaze is somewhat similar to the old, and the work as well finished, they will not resist wear and tear."

23 About the variation of the glaze of Jun kilns, there is a saying in China's ceramic world: "There's no identical Jun kiln wares." In other words, no two porcelain wares of the Jun kilns, even fired in the same kiln, were completely the same, because their glaze color was naturally formed and could not be artificially controlled. This was also in line with the highest aesthetic ideal pursued at the time. Another feature of Jun ware was the **"earthworm trail pattern,"** which refers to marks on the glaze surface which looked like the trail left by an earthworm moving through mud. This was actually a defect that occurred during the firing of Song Jun ware. The glaze layer of Jun ware was very thick and thus was prone to cracking when the temperature was still low as the kiln heated up. When the temperature moved higher, part of the glaze with lower viscosity would flow into the cracks to fill them, thus creating the earthworm trail pattern, which, like crackled glaze, was a defect formed unintentionally in the firing process but which achieved a distinctive decorative effect. Jun ware relied on unique glaze colors for decoration and was therefore in no need of any engraved, carved or stuck ornamentation.

Text Related Information

The text *Chun Yao* is adapted from chapter five of the book *Oriental Ceramic Art* (1897), written by Stephen W. Bushell, published by D. Appleton and Company, New York, and chapter eight of the book *The History of Chinese Ceramics* (2013), written by Lili Fang, published by Foreign Language Teaching and Research Publishing Co..

Selected Plates Related to the Texts

Plate 15 Basin with carved peonies and a copper rim, Ding ware. Height: 8.8 cm, diameter: 27.7 cm. Northern Song dynasty, about 1086 – 1127 A.D..

Plate 16 Ding kiln bowl, Song dynasty. Height: 9.3 cm, top diameter: 21.8 cm, base diameter: 6.6 cm, in the Palace Museum.

Plate 17　Porcelain kilns in the North, Ru kiln tripod plate, Song dynasty. Height: 3.6 cm, top diameter: 18.3 cm, distance between feet: 16.7 cm, in the Palace Museum.

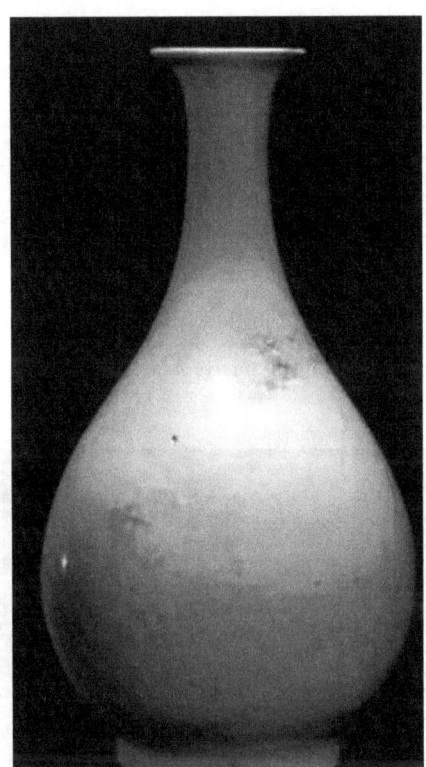

Plate 18　Blue-glazed long-neck vase, Northern Song dynasty, in the British Museum.

Unit 4 Song Dynasty Kilns in Northern China

Plate 19   Cizhou kiln "made by Zhang" black and white pillow, Song dynasty. Length: 28.3 cm, width: 19.8 cm, height: 10.5 cm, in the Palace Museum.

Plate 20   Dengfeng kiln pearl ground vase with carved pattern, Song dynasty. Height: 31.9 cm, top diameter: 7.1 cm, base diameter: 9.5 cm, in the Palace Museum.

Plate 21　Bottle with purple-splashed blue glaze, Jun ware. Height: 20 cm. Northern Song dynasty, about 1000 – 1127 A. D. .

Plate 22　Jun kiln rose-purple glaze sunflower-style flowerpot, Song dynasty. Height: 15. 8 cm, top diameter: 22. 8 cm, base diameter: 11. 5 cm, in the Palace Museum.

Aids to Comprehension

Notes

(1) R. L. Hobson　R. L. 霍布森(1872—1941)，英国知名艺术鉴赏家，20世纪上半叶英国最重要的汉学家之一，精通中国陶瓷研究，西方研究中国陶瓷的权威学者，东方陶瓷学会创始人之一。他发表了一系列有关中国陶瓷的论文和著述，代表作有《中国陶瓷史》《中国艺术》等。在现代英国的中国陶瓷收藏、展览与研究过程中，霍布森起到了重要的推动作用。基于他的陶瓷研究和艺术传播，西方对中国陶瓷和艺术有了更深入的了解。

(2) Tz'u Chou　磁州，位于河北省的南部，隶属于河北省邯郸市。磁州窑创烧于北宋，是中国北方民间瓷窑的杰出代表，也是中国北方最大的一个民窑体系。

(3) Yun shih chai pi tan　《韵石斋笔谈》，清姜绍书撰，记所见古器、书画及诸奇玩。

(4) Ju-chou　今汝州市，河南省辖县级市，位于河南省中西部，为中国汝瓷之都。

(5) Ching po tsa chih　《清波杂志》，作者周辉，是一部较为著名的宋代笔记，记载了宋代的一些名人逸事以及当时的一些典章制度、风俗、物产等。该书也保留了不少宋人的佚文、佚诗和佚词。

(6) Ching pi tsang　《清秘藏》，古代工艺美术鉴赏著作，明张应文撰。该书对器玩的辨别和收藏记叙甚详。

(7) Cho keng lu　《辍耕录》，元末明初文学家陶宗仪创作的一部笔记，主要记载宋元两朝的典章制度、史事杂录、文物科技、民俗掌故等，保存了丰富的史料。

(8) Guantai Town　观台镇，位于现今河北省磁县的西南部，是一个历史悠久且文化底蕴深厚的地方。观台镇是磁州窑瓷的重要产地之一，镇西北有古瓷窑遗址。这些遗址见证了磁州窑的辉煌历史，也为后人留下了宝贵的文化遗产。

(9) Pengcheng Town　彭城镇，坐落于河北省邯郸市，是北宋磁州窑的核心产区,与磁县观台镇共铸磁州窑辉煌。北宋时期,彭城镇陶瓷业盛极一时,成为北方首屈一指的民间瓷窑。其磁州窑遗址以独特的艺术风格和卓越的技艺闻名遐迩,是研究北宋陶瓷工艺与社会风貌的珍贵历史文物,对后世的陶瓷发展影响深远。

(10) Po wu yao lan 《博物要览》,明代谷泰撰。此书论列古器物、字画、织绣、印宝等艺术品,历来为古玩藏家、经营者、学者所重视,至今仍有很大影响。

Ceramic Terminology

(1) tear(s) "泪痕"是瓷器在烧窑过程中形成的自然现象,透明釉在高温熔融状态下因重力垂流,积釉处形成蜡泪或玻璃珠状凸起,犹如"泪痕",这是定窑白瓷的典型特征之一。

(2) hua hua (carved decoration) 刻划花,是指在尚未干透的器物表面以铁、木、竹等工具画出线状花纹,然后施釉或者直接入窑烧制。

(3) yin hua (stamped or moulded decoration) 印花,用有花纹的陶制印具,在尚未干的器物坯体上印出花纹,或用有纹样的模子制坯,直接在坯体上留下花纹,然后入窑或施釉后入窑烧制。moulded decoration,即模印贴花,亦称"贴花"或"印贴花"。其做法是用印模画出纹样,贴于坯体表面后施釉烧成。

(4) hsiu hua: The word hua (lit. flowers) is used in the general sense of "ornament." 绣花或划花

(5) Tzu: purple or dark red brown, is like most Chinese color-words, a some-what elastic term. The dictionary gives instances in which it is applied to "red sandal wood," "brown sugar," the ruby, the violet, and the peony. 紫

(6) the blue of the sky after rain 雨过天青色

(7) egg white: a white ware with a faint greenish tinge 卵白

(8) tzu run: rich and unctuous 滋润

(9) cornaline: a red, brown or white stone, used in jewellery 玉髓,玛瑙。玛瑙是一种半透明的玉髓。此处讲的是玛瑙入釉的特点。

(10) ruby red 红宝石

(11) cheng ting, may mean "engraved with a point" or "cut nails". 撑钉、支钉,用于支撑陶瓷器物的工具,特别是在宋代汝窑瓷器中较为常见。汝瓷的制作采用了一种特殊的方法,称为满釉裹足,即在器物的底部用细小的支钉支撑,使整个器物被釉完全包裹。这些支钉非常细小,以至于被形象地比喻为芝麻点。由于汝瓷采用了这种独特的烧制技术,器物底部可以清楚地见到这些由细如芝麻般的支钉形成的痕迹。

(12) pi-se ware 秘色瓷,越窑青瓷精品之一。秘色瓷是进贡朝廷的一种特制的瓷器精品,因其制作工艺秘而不宣得名。

(13) Sgraffito: refers to cutting out the ground surrounding the patterns so that the patterns appeared in relief 剔花,是一种陶瓷装饰技法,指在陶瓷坯体上刻好纹饰后,将纹饰以外的部分剔去,以凸显出设计的图案。磁州窑剔花主要分为留花剔地和留地剔花。磁州窑剔花是磁州窑特有的一种剔花装饰技法。

(14) rough blue earth 大青土,产于河北邯郸的一种半软质黏土,是磁州窑瓷器坯体的主要原料。其矿物成分主要包括高岭石、少量水白云母、石英及碳酸盐,并含有较多的植物质等有机质。大青土的塑性比白坩土好,但烧成收缩率大,容易变形。因此,在制作瓷器时,大青土通常与一定比例的白坩土混合使用,以改善其烧成性能。

(15) white engobe: a liquid white coating applied to ceramic surfaces before glazing to enhance decoration and refine the final appearance 白化妆土,可增强陶瓷表面的装饰效果,或者掩盖坯体表面的一些缺陷。白化妆土常被用于在深色或粗糙的坯体上形成对比,使最终烧制成的陶瓷表面更细腻、美观。

(16) orange-red hue 橙红色色调

(17) vermilion red 胭脂红

(18) bright onion-green 葱翠

(19) parrot green 鹦鹉绿

(20) aubergine purple 茄子紫

(21) earthworm trail pattern 蚯蚓走泥纹,钧瓷釉面上的一种特殊纹路,形似蚯蚓在泥地上爬过留下的痕迹,呈蜿蜒曲折而又长短不同的线状,有的为单线条,有的为多线条相互交叉。

Exercises

Reading Comprehension Questions

Please answer the following questions according to the text.

(1) How to distinguish Pai Ting from Tu Ting?

(2) What are the three kinds of ornament usually associated by Chinese writers with the Ting ware?

(3) What are the two mysterious peculiarities attributed to the Ju ware?

(4) What was the main raw material used for the body of Cizhou kiln ware and how did the craftsmen address its perceived inferiority?

(5) How is the "earthworm trail pattern" on Jun ware formed, and was it originally a defect or intentionally created as a decorative element?

Translation

A. Please translate the following paragraphs into Chinese, paying special attention to ceramic terminology, context and the social background involved.

(1) It was a round vessel with ear handles and three feet, and the inscription "李西涯" (Li Hsi Ya) was engraved in seal characters on the stand. During the Japanese raids in the Chia Ching period this vessel passed into the hands of one Chin Shang-pao, who sold it to T'ang, the President of Sacrifices (Tai ch'ang) of P'i-ling. T'ang, whose residence bore the romantic but chilly name of *Ning-an* (Frozen Hut), is the celebrated collector. "T'ang had many wonderful porcelains," the story runs, "when this vessel arrived, they all, without exception, made way for it. And so throughout the land when men discuss porcelains, they give the first place to T'ang's white incense vase." "T'ang, they say, did not readily allow it to be seen."

(2) The glaze was thick and of a color variously described as "approaching the blue

of the sky after rain" (i. e. like the Ch'ai ware), pale blue or green, and "egg white" which seems to imply a white ware with a faint greenish tinge. The author of the Ching pi tsang, a work of considerable repute published in 1595, gives a first-hand description of the ware: "Ju Yao I have seen. Its color is 'egg white' and its glaze is lustrous and thick like massed lard. In the glaze appear faint 'palm eye' markings like crabs' claws. Specimens with sesamum designs (lit. flowers), finely and minutely engraved on the bottom, are genuine. As compared with Kuan Yao in material and make, it is more rich and unctuous (tzu run)."

B. Please translate the following paragraph into English.

宋代陶瓷的昌盛得益于城市经济与文化的发展，而宋代陶瓷中的那种特殊的审美趣味则源于宋代文人的美学追求。由于都市生活的发展和商业的繁荣，瓷器成了当时人们生活中的主要器皿，同时也是当时文人品茶饮酒的用具和把玩物。宋瓷作为这些文人饮茶赋诗时的道具、风花雪月时把玩的对象，寄托了他们的情思、他们的理想、他们的品位、他们的审美标准。在历史上，中国的食器从来就不仅仅是一种食器，它是一种礼仪与制度的体现，也是一个人的身份、地位和权力的象征。在夏、商、周时，它是礼器的重要组成部分。到官窑建立以后，它更确定了作为礼器的地位，具有一种文化意义上的象征性。由于这种文化意义上的象征性的引导，也就有了"器以载道"的说法。"器"是具体的物质，是形而下的有用之物，而"道"则是器所包含的一种文化内涵，是属于形而上的精神性的东西。这一点在宋瓷的审美中得到了更清晰的体现。

Unit 5 Song Dynasty Kilns in Southern China

The Song dynasty witnessed a flourishing era of porcelain production in southern China, with numerous kilns dotting the landscape. These kilns collectively showcased advanced firing techniques and refined aesthetics, contributing significantly to the development of Chinese ceramics. The following delves into the rich history and exquisite craftsmanship of three renowned kilns during the Song dynasty: the Guan kilns of the Southern Song, the Longquan Yao, and the Jingdezhen kilns. The Guan kilns, showcase a continuation of the Bianjing Guan kilns' legacy, producing celadon porcelain prized for its jade-like glaze and exquisite designs. The Longquan Yao, originating from Longquan region, is renowned for its crackled celadon and green porcelain, which captured the imagination of collectors worldwide. Lastly, the Jingdezhen kilns, rising to prominence in the Five dynasties and maturing during the Song, revolutionized porcelain-making with its qingbai ware, hailed for its bluish-white glaze resembling ice and jade. This exploration sheds light on the technological advancements, artistic expressions, and cultural exchanges fostered by these kilns, showcasing the pinnacle of Chinese ceramic art during the Song dynasty.

Text

The Guan Kilns

Lili Fang

1 Song official kilns included Northern Song Guan kilns and Southern Song Guan kilns, the former located in Bianjing and the latter established in Hangzhou by the Song royal family after they moved south. To facilitate production management and product transportation, the royal family built official kilns near the capital. This part mainly introduces the Guan kilns of the Southern Song.

2 The Guan kilns of the Southern Song dynasty were newly set up in Hangzhou after the Northern Song ended and Emperor Gaozong fled to the south. At that time Lin'an (now Hangzhou) was the center of politics, culture and the economy as well as the largest consumer city. There were also *Mingzhou* (now Ningbo), *Yuezhou* (now Shaoxing), Wenzhou, *Wuzhou* (now Jinhua), and *Chuzhou* (now Lishui). These cities and towns, with long histories and an advanced porcelain industry, were both production centers and markets and distribution centers of porcelain. Yue ware, Wuzhou kiln ware and Longquan celadon had a long-established fame for their unique features. The **Jingkang Incident** marked the end of the Northern Song regime and amidst the war, the most famous kilns on the central plains were destroyed. The Song royal family, led by Zhao Gou, fled southeast, made Lin'an his capital and set up the Southern Song dynasty. The Southern Song court established the Xiuneisi Guan kiln on Phoenix Mountain, Hangzhou, and a new Guan kiln near Jiaotan in the Bagua Cave, Tortoise Mountain. In the Xiuneisi kiln and the new Guan kiln the best craftsmen from the north and south gathered to fire celadon required by the court and dignitaries. These kilns were a continuation of the Bianjing Guan kilns of the Northern Song dynasty and were historically called the Guan kilns of the Southern Song. The earliest record of the Guan kilns of the Southern Song is found in the ***Notes in Tan Study*** by Ye Zhi of the Southern Song: "During the Zhenghe years the capital established its own kilns to fire porcelain, called the Guan kilns. After the royal family crossed the Yangtze River, Shao Chengzhang proposed

establishing the Shao Bureau. A kiln was set up in Xiuneisi to make celadon, called the Nei kiln, inheriting the legacy of the former capital. Clay was finely washed to be made into models. The products, extremely exquisite with clear glaze, were favored by the people. Later a new kiln was set up in Jiaotan, which was quite different from the old kiln. The others, such as the Wuni, Yuhang, and Xu kilns, could not be compared with the Guan kilns. As for the old Yue kilns, they no longer existed. " This description is the earliest written record discovered to date of the history of Southern Song's Guan kilns and has been used as textual research materials by subsequent researchers.

3 According to the text, the Guan kilns were set up back in the Northern Song dynasty and the Guan kilns of the Southern Song dynasty inherited their technology and production methods. The text classified Southern Song's Guan kilns into "Xiuneisi" and "Jiaotan," the former called "the Nei kiln" and the latter "the new kiln, " and explicitly pointed out that the new kiln "was quite different from the old kiln. " It is mentioned that the Xiuneisi kiln was also named the "Nei kiln, " that it used finely washed clay to make models, and produced extremely standard products that were cherished by the people for their clear and transparent glaze. According to the *Ge Gu Yao Lun*, an early Ming manual on ceramics by Cao Zhao, "Products fired by the Song dynasty's Xiuneisi (Guan kiln) were made from fine clay, had a blue color with a pinkness of differing shades, and featured a crab claw pattern, purple rim and iron feet. The wares of fine color were similar to Ru ware. Those made from black clay were called Wuni wares. Fake ones were fired by the Longquan kilns and had no pattern. " Here the characteristics of the porcelain fired by the Xiuneisi kiln were described in detail. As for the location of the kiln, Gao Lian of the Ming dynasty wrote in his book **Zun Sheng Ba Jian** that "the Guan kiln was at the foot of Phoenix mountain, " yet for many years people only found a lot of scattered porcelain shards and fragments of kiln furniture in the area and had not really discovered the exact location of the kiln site.

4 In September 1996, the Tiger Cave kiln site near the Southern Song imperial city site on Phoenix mountain was discovered by chance when flooding exposed large

numbers of celadon shards and kiln furniture. In November that year, the Hangzhou Institute of Cultural Relics and Archaeology conducted a one-month-long archaeological survey of the site and named it the Tiger Cave kiln site following the place's name as used by the local people. The geographical location of the kiln site is very special because it coincides perfectly with the location of the Southern Song Xiuneisi Guan kiln as recorded. From then until 2001, archaeological departments comprehensively inspected the celadon kiln site which spanned the Southern Song and Yuan dynasties with three excavation and investigation works. The large number of site remains and physical materials excavated from the Southern Song dynasty strata showed conclusively that the site was once an important part of the Guan kilns of the Southern Song dynasty.

5 During excavation, the Hangzhou Institute of Cultural Relics and Archaeology unearthed a large number of Guan kiln porcelain shards and kiln tools. The main varieties were bowls, plates, cups, jars, saucers, pots, washers, lamps, chopstick shelves and other daily utensils, as well as decorative wares such as *gu* (goblets with broad lips, long narrow stems, and flared bases), **cong-style bottles**, incense burners, **pedestals**, and flowerpots. Body color might be incense ash color, dark gray, purple, or black. Purple rims and iron feet were common. Thick bodies and thick glaze layers dominated while thin bodies or thin glaze layers were rare. Glaze color was mainly greenish blue and beige yellow, as well as emerald green, grayish blue, light purple, and yellow. Most glazes had ice cracks. Some of the porcelain shards were found to have the words "Xiuneisi" and "Guan kiln" written in brown under the glaze.

6 Following the Xiuneisi Guan kiln, the Southern Song dynasty built a second Guan kiln called the Jiaotan Guan kiln, whose site is located in the area of Tortoise mountain in the southern suburbs of today's Hangzhou. The site was discovered at the beginning of the twentieth century and porcelain shards had subsequently constantly flowed into the antique markets of Hangzhou and Shanghai and thus attracted the attention of ceramics researchers at home and abroad. In the 1950s the Zhejiang Provincial Cultural Relics Management Committee carried out a small-scale trial

excavation of the site, cleaned up a dragon kiln, and excavated some of the shard accumulations beside the kiln. Then a Lin'an archaeological team officially excavated the site in 1985 and conducted supplementary excavations in 1988 to support the building of the Southern Song Guan Kiln Museum. These excavations offered a clear understanding of the Southern Song Jiaotan Guan kiln. Of the unearthed porcelain shards, 23 varieties of wares were complete or restored. In addition to daily utensils which were unearthed in the largest number such as bowls, plates, bottles, jars, pots, and basins, there were also a number of decorative objects such as gu, cong-style bottles, incense burners, pedestals, and flowerpots. The decorative objects were mostly for ceremonial use and imitations of bronze and jade wares of the Shang, Zhou, Qin and Han dynasties. Wares can be classified into thick and thin based on body thickness, and glaze color was mainly greenish blue and beige yellow, though shades of yellowish-brown and earthy yellow were also common. Daily use utensils accounted for the largest number of the wares unearthed, yet among the Song Guan wares collected by the Palace Museum and the Taibei Palace Museum, daily use utensils are fewer while display ceramics account for the majority, which is probably because the former suffered a high rate of damage while the latter enjoyed better preservation. This also shows the important position of display ceramics as ceremonial wares in the production of the Southern Song Guan kilns. Southern Song Guan celadon excelled mainly through its jade-like mellow glaze color, its archaic, simple and dignified shape design, along with its crack pattern, and therefore it did not require any rich decorative themes. Besides common plant and animal patterns, there were also various **string patterns**, Eight Trigrams patterns, **yunlei patterns**, square spirals, double circles, and bump patterns. Commonly used decorative techniques were engraving, molding, **stacking**, kneading, **hollowing out** and **openwork**; engraving was mostly used on daily use utensils such as bowls and plates, molding was used on a wide range of varieties, stacking was mostly used on wares imitating ancient vessels such as bottles, pots, stoves and gu, and openwork was mostly used on lids, pedestals and stoves. The rich decorative techniques, the improvement in firing technology and the diversity of kiln tools for loading and firing on the one hand reflected the attention the Southern Song Guan kilns were receiving in terms of production technology and on the other hand explained the reasons behind the excellent quality of their products.

7 Guan kiln craftsmen who migrated south in the Southern Song dynasty brought technology to Hangzhou. The conditions in the south for porcelain firing were relatively superior to those in the north. At that time the south already had a solid foundation with the Yue kilns, secret-color ware and the Longquan kilns and it would have been easy and quick for the south to learn to fire Guan wares. In the early years of the Southern Song dynasty the products still demonstrated a heavy influence from the Northern Song Guan kilns, mainly in their thin body, thin glaze layer and the use of prop nails for firing. For instance, the samples unearthed at the Jiaotan kiln site had a thin, fine body which could be light gray, gray or deep gray. The outer surface had a thin layer of greenish blue, bluish gray or beige yellow glaze applied. The commonality of beige yellow glaze indicates that the porcelain firing technology of potters in this period was still in the exploratory stage. The circular feet of bowls, plates and washers were high and flared. Nails were used to prop the wares during firing, leaving needle-like marks on the glaze of the outer bottom.

8 As the Southern Song regime consolidated its power, the imperial court required higher quality porcelain. Consequently, the technology of the Guan kilns was constantly improved. The technique of multiple firing and multiple glazing was adopted to achieve an improved jade texture of celadon, and the mouth and belly wall of wares, especially of small wares such as bowls, plates, washers and saucers, was thinned down to make the wares more lightweight and graceful. The greatly thick-ended glaze layer also prompted kiln workers to reform kiln tools, such as replacing prop nails with **cushion pads** to avoid the nails getting stuck to the thick glaze during firing and rendering the products defective. The adoption of cushion pads allowed the circular foot of wares to take on the function of bearing the weight of the ware, prompting previous flared circular feet to be gradually replaced by straight ones.

9 The Southern Song Guan ware not only inherited the simple yet dignified shape

design and thick glaze texture of famous northern wares such as the Northern Song Bianjing Guan and Ru wares, but also assimilated the thin body combined with thick glaze, bright, clear glaze surface and exquisite shapes of famous southern wares such as Yue and Longquan wares. Cao Zhao of the Ming dynasty described Southern Song Guan ware in his *Ge Gu Yao Lun* as: "Products fired by Song dynasty's *Xiuneisi* (Guan kiln) were made from fine clay, had blue color with pinkness of different shades, and featured crab claw patterns, purple rims and iron feet."

10 The characteristics of Southern Song Guan ware could be summarized as **"purple rim and iron feet,"** "greenish blue glaze," "ice crack patterns" and "thin body and thick glaze." Of the seven major glaze colors of Guan ware, namely, beeswax yellow, pale blue, greenish blue, light greenish blue, deep greenish blue, crab color and putty color, greenish blue was the finest. Repeated polishing and firing of the greenware achieved an extremely thin body while the glaze thickness was more than three times that of ordinary ceramic glaze. It can be seen that the attraction of the Southern Song Guan celadon lay not in its decoration, but mainly in its jade-like, dignified, elegant and mysterious natural beauty.

Text Related Information

This text is adapted from chapter eight of the book *The History of Chinese Ceramics* (2013), written by Lili Fang, published by Foreign Language Teaching and Research Press, Beijing.

Lung-chuan Yao

Stephen W. Bushell

11 The Lung-chuan Yao is the porcelain that used to be made at Lung-chuan-hsien, in the prefecture Chu-chou-fu, in the southern part of the province of Chekiang. During the early part of the Song dynasty the factory was at **Liu-tien**, some twenty miles distant from the walled city of Lung-chuan, and under its

jurisdiction. Two brothers named Chang, who are said to have lived here in the twelfth century of our era, are celebrated for their productions. The elder, called for that reason Chang Sheng yi, introduced a new glaze, distinguished by its crackled texture, which became known as Ko Yao, or the "Elder Brother's Porcelain." "Chang Secundus," Chang Sheng erh, fabricated his ware on the old lines, only improving the luster and color of the green glaze, so that his productions continued to be called by the old name of Lung-chuan Yao.

12　These potteries furnished the main source of the famous old celadon and crackled porcelain, which were exported at this time from China to all parts of Asia, as well as to the eastern and northern coasts of Africa. They constitute the Ching Tsi, the "green porcelain," par excellence of the Chinese, and are well known to the Japanese, who esteem them very highly by the same name, which they pronounce **Seiji**. During the Song dynasty there was considerable commercial intercourse by sea between China and the Mohammedan countries, and we read in both Arabian and Chinese books of the time of "green porcelain" as one of the articles of trade. The Chinese describe it as carried as far as Zanzibar, which they call Tsangpa, and are curiously confirmed by the discovery there in some old ruins, during Sir John Kinks residence as H. B. M. consul-general, of a quantity of celadon vessels, principally in fragments, mixed with Chinese coins of the Song dynasty.

13　The Arabs and Persians call this peculiar porcelain martabani, and value it very highly from its fancied property of detecting poisoned food by changing color. The name comes from Martaban, one of the states of ancient **Siam**; and Prof. Karabacek, of Vienna, has lately tried to prove that it is not Chinese, basing his theory mainly upon a passage quoted from the encyclopedist Hadji Khalifa, who died in 1658, that "the precious magnificent celadon dishes and other vessels seen in his time were manufactured and exported at Martaban, in Pegu." But there is no evidence that porcelain was ever made at Maulmain (Martarban), Rangoon, or elsewhere in Burma. Others have attributed it, with little success, either to Persia or to Egypt, because so much has been discovered there, but neither of these countries produced true porcelain, although they excelled in the decoration of faience. An

Arab manuscript in the Bibliotheque National at Paris, treating of the life and exploits of Saladin, mentions that this emir presented in the year of 1171 forty pieces of this kind of Chinese porcelain to Nuredin.

14　Marco Polo, after his travels in China in the thirteenth century, seems to have been the first in Europe to use the name to describe this product of the far East. It had probably been applied previously only to shells and Marco Polo applied the same term to the cowries which he found used as money in eastern countries.

15　The crusades were apparently the earliest means of introduction of specimens of this ware to the west. Dr. Graesse relates that the most ancient piece in the Dresden Museum was brought by a crusader from Palestine, perhaps it came through Egypt. A present of porcelain vases was sent in 1487 by the Sultan of Egypt to **Lorenzo de Medici**, and it is mentioned about the same time in the maritime laws of Barcelona as one of the articles imported into Spain from Egypt.

16　The glaze of the Lung-chuan porcelain is of a monochrome green color, varying from bright grass-green, the tint of the Chinese olive, a species of canarium, through lighter inter-mediate shades to palest sea-green. The term celadon is well known to collectors as applied to these different shades of color. Celadon was the name of the hero of the popular novel *L'Astrie*, written by Honore d'Urfe in the seventeenth century, who used to appear on the stage dressed in clothes of a kind of sea-green hue of a gray or bluish tint. This shade became fashionable, and the name was borrowed to describe a similar shade in the color of Chinese porcelain. This peculiar shade, however, is specially characteristic of the Lung-chuan porcelain of the Ming period, made in the city of Chu-chou-fu, to which place the manufactory was removed early in the Ming dynasty. It was here that the characteristic large dishes were made marked underneath with a ferruginous ring, showing the portion of the paste left unglazed, so as not to adhere to the support in the kiln. The older pieces attributed to the Song dynasty are completely covered with glaze under the foot, and are generally of a more decided grass-green color, approaching the emerald-green tint of jadeite, which seems to have been the effect especially aimed at. The decoration was either worked in relief or engraved in the paste, and its effect was enhanced by the different shades of color produced by the varying depth of the glaze. The vessels are often fluted or **ribbed** and with wavy or foliated rims; some have a peony or a lotus blossom, fish or dragons, **sprays of flowers** or geometrical patterns **etched** in the paste. Others have a pair of fishes worked in relief in the bottom, or a pair of rings attached outside to handles.

Text Related Information

This passage is adapted from chapter five of the book *Oriental Ceramic Art* (1897), written by Stephen W. Bushell, published by D. Appleton and Company, New York.

The Jingdezhen Kilns

Lili Fang

17 The Jingdezhen kilns started production in the Five dynasties and developed a mature porcelain-making technology by the Song dynasty. In the Five dynasties when the Jingdezhen kilns arose, there emerged the prosperous scene of every village hosting a kiln and every family molding pottery. Large numbers of Song dynasty ceramics deposits have been found in ancient kiln sites at many places in the suburbs of Jingdezhen, such as Liujiawan, Nanshijie, Zhuxi, Sitian, Jiantian, Yangmeiting, Hutian, Shihuwan, and Huangnitou. This shows that during the Song dynasty folk kilns in the area were gradually concentrating around what is now urban Jingdezhen and that the porcelain-making technology had been greatly improved. In fact, the name "Jingdezhen" itself came into existence only during this period. Before the Song dynasty the place was named Xinping or "Changnan," and was renamed Jingdezhen ("Jingde town") during the Jingde era of Emperor Zhenzong's reign in the Northern Song dynasty. *The Comprehensive History of Jiangxi* records: "During the Jingde years of Song the town was established, and officials were ordered to make porcelain for the capital to be used by the government. Potters were ordered to write

'Jingde' on the wares." Meanwhile, Lan Pu in the fifth volume "Examinations of Kilns of Past dynasties" of his *Accounts of Ceramics at Jingdezhen* notes that: "all called it Jingdezhen ware and 'Changnan' as a diminutive name."

18 From Emperor Zhenzong's reign to the end of the dynasty the Song court established and maintained tax administrators in Jingdezhen called "Town Supervisors" and appointed officials to preside over the production of the folk kilns. Thus, the porcelain industry of Jingdezhen in the Song dynasty was a private handicraft industry and its development also reflected the development of commodity production as a whole in the Song dynasty.

19 Regarding the characteristics of Jingdezhen's porcelain in the Song dynasty, it is recorded in the fifth volume "Examinations of Kilns of Past Dynasties" of *Accounts of Ceramics at Jingdezhen* that: "Jingdezhen selected white clay to mold the body, the texture is thin and fine, the color is mellow and lustrous… the ware is exceedingly bright and attractive. It became popular and was held as an example in the country." In *Records of Porcelain Making* Jiang Qi recalls Jingdezhen ware in the Northern Song dynasty as being: "all praised to be like 'beautiful jade'."

20 Jingdezhen's main porcelain products at the time were qingbai wares with a thin body, clear glaze and jade-like color. They were renowned as the champion of all kilns for their pure **texture** and fine workmanship and became a precious product that represented the top porcelain-making level of the Song dynasty. Under their influence, many kilns at the time in today's Fujian, Guangdong, Sichuan, Zhejiang, Anhui, Hubei, Yunnan, Guangxi and other provinces and prefectures all started imitating Jingdezhen in firing qingbai porcelain. The impact of the Jingdezhen kilns was so far-reaching that a "qingbai porcelain kiln system" was formed and became an important chapter in the history of China's ceramics. Within the Jiangxi kilns imitating qingbai porcelain were the Jizhou and Baishe kilns in Nanfeng, the Qilizhen kilns in Ganzhou, as well as the Ningdu, Leping, and Jing'an kilns.

21 In the Five dynasties, what Jingdezhen produced was celadon and white porcelain. Qingbai porcelain had not appeared at that time. The general development trend of China's porcelain kilns was to inherit and innovate on the basis of the

traditional types of the previous period, yet Jingdezhen was an exception. By the Song dynasty, the celadon and white porcelain that were popular in Jingdezhen back in the Five dynasties disappeared and were replaced by qingbai porcelain. The reason behind this, as posited by **Feng Xianming**, was probably that craftsmen in Jingdezhen intentionally fired qingbai porcelain with a jade-like texture by ingeniously utilizing excellent local materials to meet consumer needs because jade had always been a rare commodity monopolized by the ruling class and bluish white jade, while desirable, was even more rare and hard to come by. It was because this porcelain type was created to imitate the color feature of bluish white jade that its glaze color was between blue and white. Hence, "*qingbai* (bluish white) porcelain" was used to refer to whitish blue ware or bluish white ware. It was called "*ying* (影) *qing* (shadow blue)" by the folks, and sometimes "*ying* (映) *qing* (reflected blue)," "*yin* (隐) *qing* (subtle blue)," and "*zhao qing* (covered blue)." The qingbai porcelain produced by Jingdezhen in the Song dynasty had a fine, white, delicate body, a bluish white glaze color resembling ice and jade, a color so bright and clear that it was able to show reflections. Mellow and crystalline, elegant and refined, such porcelain really deserved the praise "like beautiful jade" as recorded in literature.

22 Almost all of Jingdezhen's qingbai porcelain kilns in the Song dynasty were distributed in the Donghe and Nanhe river basins, with the majority concentrated in the Nanhe river basin where porcelain stones were produced.

23 Representative kilns in the early Song dynasty were in Huangnitou, Xianghu, Shihuwan, Hutian, and elsewhere. Qingbai porcelain of this period featured a relatively thick body and obvious signs of being **"shaped from wet clay."** Pots and similar varieties at the time were shaped by throwing and processed while the clay was wet. The forming of bowls and plates involved three procedures, namely throwing, internal mold casting and **foot embellishing**. As to the method of loading wares in the kiln, the Five dynasties practice of pile firing was abandoned. Wares were instead protected by saggars and multiple cushion pads, and fired face up, a method that was called **"face-upward firing."** Owing to insufficient mastery of the

reducing flame, the glaze color of products of the period lacked green notes and was more gray or light beige yellow. The shape designs were similar to those of the Five dynasties. Most bowls and small cups maintained the **rolled rim** or **sunflower rim** of the Five dynasties and featured a high circular foot and **slanting belly wall**, or **short circular foot** and funnel shape. Pots were melon-shaped and still had signs of imitating metal ware. Pots unearthed from early tombs all had a lid topped with a lion figurine. Almost no ware had a decorative pattern on its surface. Bowls, small cups, etc. were mostly fired using the "face-upward firing" method. In general, qingbai wares from this period had monotonous shapes and had not developed the characteristics of *yingqing* (shadow blue) glaze. Yet, compared with the regular celadon and daily use porcelain of the time, they show a fairly high level of craftsmanship.

24 By the mid-Song dynasty, obvious **equidistant spinning patterns** appeared on vessels, indicating that the wares were preliminarily shaped by throwing and then, after becoming **leather hard**, trimmed with an iron blade while being spun. The resultant utensils were thus much thinner, finer and neater than those from the Five dynasties or early Song dynasty. Due to the special feature (**poor plasticity**) of the raw material used by Jingdezhen porcelain, trimming, also called turning, was widely used to form the wares. By the middle of the Song dynasty, a forming approach where turning played the major role, while throwing was supplementary, emerged in Jingdezhen, setting it apart from northern kilns. Nevertheless, the most outstanding achievement of Jingdezhen potters in this period was the skillful mastery of the **strong reducing flame** which not only realized a mellow jade-like bluish white color, but also improved the density and transparency of the ware body. The glaze color of most wares turned from the grayish blue or fired rice yellow of the early Song dynasty to bluish green. Thus, the glaze layer of this period began to display the characteristics of typical yingqing glazed porcelain. Manufactured in the greatest numbers were bowls and plates, which were of many varieties. There were also vases, incense burners, porcelain pillows, decorative sculptures and other artworks, which were elegantly shaped, thin walled, and decorated with carving and painting.

The decorative patterns had graceful lines and strong artistry with such themes as water ripples, grass, peony, lotus, fish, and phoenix bird. Of the kilns producing yingqing porcelain in different places, Hutian and Xianghu achieved the highest technological level. Their products had a fine body texture, thin wall and attractive shape design, and both their transparency and hardness reached the standard of modern hard porcelain.

25 In the late Song dynasty (the Southern Song), Jingdezhen began to use the kiln tools combined with cushion pads which had been invented by the Ding kilns to hold bowls and plates with an unglazed rim for firing. Compared with multi-layer cushion bowls, such kiln tools did not require saggars and were able to accommodate multiple **greenware** of the same specifications. Compared with face-upward firing saggars, they could increase the vertical holding capacity of the kiln space by four to six times and thus save 75% of the fuel, and in the meantime could prevent product deformation. Yet using this tool would leave an unglazed rim on the wares, thus affecting their practical value. In addition, the fact that the tool was made of greenware and could only be used once meant that during firing its usage increased moisture in the kiln space which was not easy to eliminate. This in turn slowed down the temperature rise in the kiln and easily led to defects such as damp yellowness of the glaze. Due to the popularity of **inverted firing**, engraved or carved patterns on **round wares** were reduced and gradually replaced by printed decoration featuring complicated and over-elaborate patterns, which were rigid compared to engraved patterns but which facilitated production. Popular pots and bottles in later periods were also mostly imprinted with external molds, resulting in simple shapes and a rigid appearance. Daily use bowls and small cups mostly had **unglazed rims** and the shape design changed from the previous style featuring lighter tops and heavier bottoms to the present one featuring thicker mouths and rims and thinner middle and lower parts. The circular foot became shorter than the early and middle periods. Some wares only had a trace of the circular foot left. The structure of bowl walls was also designed to suit the requirements of inverted firing, that is, to increase the stacking density in the kiln chamber. The adoption of inverted firing to make

yingqing porcelain in the Southern Song dynasty was on the one hand influenced by the Ding kilns in pursuit of better profits and reduced costs, and on the other hand due to a change in raw materials. **Liu Xinyuan** believed that only one kind of raw material was used for making porcelain in the Song dynasty, namely, porcelain stone, which could be classified into upper layer porcelain stone and lower layer, the former being more suited for firing at high temperature. When the upper layer porcelain stone, which was easier to mine yet of limited reserves, was exhausted in the Northern Song dynasty, potters of the Southern Song dynasty started using middle and lower layer porcelain stone, a material that was prone to deformation. To counter possible deformation, potters placed bowls made of middle and lower layer stone upside down on cushion rings so that the bowls were dome-shaped and less prone to deformation during firing. Yet as a result of this practice almost all yingqing wares of the Southern Song dynasty ended up with an unglazed rim.

26 Compared with the Northern Song dynasty the porcelain manufacturing of the Southern Song deteriorated, the reasons for which, apart from the raw material issue mentioned above, also included social unrest and the harsh tax system, which dealt a heavy blow to the porcelain industry in Jingdezhen. Despite this setback, to provide high-quality porcelain for the imperial palace and the nobles, Jingdezhen retained the firing technologies and craftsmanship that had been skillfully mastered during the Northern Song dynasty. Most decorations were engraved or carved and pursued the effect of being "exceedingly bright and beautiful," which offered later generations a glimpse into the high craft level of Jingdezhen folk kilns in the middle of the Southern Song dynasty. In fact, by the latter part of the Southern Song dynasty, Jingdezhen had almost become the largest kiln cluster in China in terms of technology, production scale and commodity circulation. Following the mid-Song dynasty, the Jingdezhen folk kilns experienced a development process from the primary to the advanced, that is, they gradually broke free from a reliance on agriculture to become an independent handicraft industry with a complex division of labor and cooperation. Their large production scale, wide sales coverage, and high technological and artistic achievements won the Jingdezhen kilns a well-deserved fame equal to the Ding and

Longquan kilns following the celebrated "Ru, Guan, Ge and Jun" kilns. The excellent raw material of Jingdezhen resulted in the high whiteness and transparency of the qingbai porcelain body which was close to the whiteness and **transparency** of modern fine porcelain. Jingdezhen qingbai ware from the Song dynasty was the first in China to develop porcelain from transparent glaze to semi-transparent body, which was the third leap in the development of China's porcelain. This leap laid the foundation for the development of the porcelain industry in the Yuan dynasty.

Text Related Information

This text is adapted from chapter eight of the book *The History of Chinese Ceramics* (2013), written by Lili Fang, published by Foreign Language Teaching and Research Press, Beijing.

Selected Plates Related to the Texts

Plate 23 Hexagonal basin with six feet, Guan ware. Height: 8.5 cm, diameter: 21.5 cm. Southern Song dynasty.

Plate 24 Brush-washer with celadon glaze, Longquan ware. Height: 5.3 cm, diameter: 17 cm. Southern Song dynasty.

Plate 25 Pair of funerary urns with celadon glaze, Longquan ware. Height: about 25 cm. Southern Song dynasty.

Unit 5 Song Dynasty Kilns in Southern China

Plate 26 Porcelain figure with qingbai glaze. Southern Song dynasty.

Plate 27 Qingbai-glazed lidded bowl with lotus petal pattern, Song dynasty, in the Metropolitan Museum of Art in the USA.

Aids to Comprehension

Notes

（1）Jingkang Incident　靖康之难。靖康二年，金军攻破东京（今开封），掠走徽宗、钦宗和妃嫔、朝臣等 3000 多人以及城中的全部财物。北宋灭亡。

（2）Notes in Tan Study　《坦斋笔衡》，为叶寘所写，是非常重要的文献资料。叶寘为南宋人，所遗资料为当代人记当代事的文献，因此被现代宋瓷研究者称为最权威的史料。

（3）Zun Sheng Ba Jian　《遵生八笺》，是一部内容广博又实用的养生专著，为明代高濂所撰。全书分为清修妙论笺、四时调摄笺、起居安乐笺、延年却病笺、燕闲清赏笺、饮馔服食笺、灵秘丹药笺、尘外遐举笺等八笺，遍涉生活起居、衣食住行、吃喝玩乐、出行郊游、古玩鉴赏等，其中"燕闲清赏笺"，有三篇关于宋代窑器的文章：《论定窑》《论诸品窑器》《论饶器新窑古窑》。

（4）Liu-tien　琉田，距龙泉市区 40 千米，位于山清水秀的琉华山下。

（5）Seiji　青瓷

（6）Siam　暹罗，泰国的古称。

（7）Lorenzo de Medici　洛伦佐·德·美第奇（1449—1492），意大利政治家、外交家、艺术家，同时也是文艺复兴时期佛罗伦萨的实际统治者。他的统治恰逢意大利文艺复兴的繁荣时期，这在很大程度上要归功于他对艺术、文化和哲学的赞助。美第奇家族是佛罗伦萨 15 世纪至 18 世纪中期在欧洲拥有强大势力的名门望族。他们的艺术赞助行为对意大利文艺复兴运动产生了巨大助推作用，对中国陶瓷的收藏与仿制则加速了中国陶瓷在意大利乃至整个欧洲的传播。

（8）Accounts of Ceramics at Jingdezhen　《景德镇陶录》，清景德镇人蓝浦著，其弟子郑廷桂增补，共十卷。该书对景德镇窑瓷的历代沿革有较深入而全面的描述，对清代的情况所记尤详，是研究景德镇瓷器史的重要依据。

（9）Records of Porcelain Making　《陶记》，记录景德镇窑瓷器生产的宋代文献，蒋祈著。书中翔实地记载了景德镇瓷原料产地、胎釉制备、成型、装饰、焙烧、制匣等方面的内容,并准确描述了当时的陶瓷市场、地方习俗、税收制度、瓷器品类和景德镇瓷业内部的精细分工。《陶记》是研究中国陶瓷史、手工业史、科技史的珍贵文献。

(10) Feng Xianming 冯先铭,中国古陶瓷专家,主要从事中国古陶瓷的研究、整理、收购、编目、陈列以及古窑的调查和鉴定等工作。出版《龙泉青瓷》《定窑》《青白瓷》《中国古陶瓷论文集》《中国古陶瓷文献集释》等专著,主编《中国陶瓷史》等书。

(11) Liu Xinyuan 刘新园,中国陶瓷史学家、陶瓷考古学家。他为中国陶瓷史、景德镇官窑和湖田窑、高岭土、元青花、蒋祈《陶记》等领域的研究均作出了杰出贡献。

Ceramic Terminology

(1) gu: goblets with broad lips, long narrow stems, and flared bases 觚,喇叭形口,细腰,高圈足。宋代盛行以青铜觚为样本烧制瓷觚,作观赏瓷或花器用。

(2) cong-style bottle: cong-style bottle is based on the shape of jade Cong of Liangzhu Culture in the Neolithic age. Cong, with a square cylindrical shape, was the earliest jade ware with a round hole inside. It is a ritual ware in the Neolithic period, called Jade Cong. The type of bottles has round mouth, short neck, square cylinder, long body, and the size of the circle foot, mouth and foot are similar. Some bodies are decorated with raised horizontal lines on all sides. 琮式瓶。其器型是圆口,短颈,方柱形长身,圈足,口、足大小相若。有的器身四面有凸起的横线装饰。琮,内有圆孔的玉器,是新石器时期的礼器,也称为玉琮。瓷质琮式瓶沿袭了玉琮的基本形制,只是玉器内圆外方,上下通透,而瓷质琮式瓶加了圈足和底,演化为一种瓶。

(3) pedestal: the base on which something such as a statue stands. 器座,一种用于支撑器物的附属品,其主要功能是配合器物,起到稳定和陪衬的作用。

(4) string pattern 弦纹,是古代陶器及青铜器装饰的纹样之一。弦纹是刻划出的单一的线条或若干平行的线条,排列在器物的颈、肩、腹、胫等部位,是古器物上最简单的传统纹饰。

（5）yunlei pattern　云雷纹，陶瓷装饰的一种原始纹样，图案呈圆弧形卷曲或方折形的回旋线条。圆弧形的称云纹，方折形的称雷纹，云雷纹是两者的统称。

（6）stacking　堆贴，又称"塑贴""堆塑"，是将印出或塑出的立体状纹饰贴附在陶瓷的坯体上，然后罩釉烧制而成的一种陶瓷装饰技法。

（7）hollowing out　镂空，即在器物坯体未干时雕刻出装饰花纹，然后将这些部分镂空的纹样直接入窑烧制或施釉后入窑烧制。

（8）openwork　透雕，介于圆雕和浮雕之间的一种雕塑，即在浮雕的基础上，镂空其背景部分。有的单面雕，有的双面雕。一般有边框的亦称镂空花板。

（9）cushion pad　垫饼，烧瓷器的垫烧工具，多用粗耐火黏土或高岭土制作，因形似饼，故称垫饼。

（10）purple rim and iron feet　紫口铁足

（11）rib: bulging veins　出筋，即制胎时故意在某些地方做条状突起，使该处的釉比器身其他地方薄，烧成后釉色显得浅淡，这种工艺常见于南宋龙泉窑青瓷器上。

（12）sprays of flowers　折枝花纹，与之相对的是缠枝花纹（scrolls of flowers）。

（13）etch: a pattern or a line is etched into the surface of ceramic wares by means of some sharp tools.　凿刻

（14）texture　质地

（15）shaped from wet clay: Wares were shaped by throwing and processed while the clay was wet.　湿泥定型

（16）foot embellishing: foot trimming　修足，指对瓷器底部的处理过程，包括足部（即底部）的设计和装饰。

（17）face-upward firing　仰烧，瓷器装烧方法之一。坯件仰放在垫饼上后放入匣钵中，再把匣钵一个个堆垛起来入窑焙烧，这种装烧瓷器的方法，也称正烧。正烧的器物口沿都有釉，底足露胎。

（18）rolled rim　唇口，陶瓷容器口部形式之一。器口边沿凸起一道厚边，线条浑圆似嘴唇，故名。

（19）sunflower rim　葵口，陶瓷碗、盘花口形式之一。圆形器口做等分的连弧花瓣状,形似秋葵花,故名。

(20) slanting belly wall　斜腹壁

(21) short circular foot　矮圈足

(22) equidistant spinning patterns　等距旋纹

(23) leather hard　半干状态

(24) poor plasticity: low plasticity　可塑性低

(25) strong reducing flame　强还原火焰,即在燃烧过程中,由于氧气供应不足,燃烧不充分,燃烧产物中含有一氧化碳等还原性气体,火焰中没有或含有极少量的氧分子。这种还原性火焰,在瓷器的烧制过程中有特殊的作用。

(26) greenware　生坯、瓷土、耐火材料等经加工、成形、干燥但未烧成的半成品。

(27) inverted firing　覆烧,即将碗、盘等器皿反扣着焙烧的一种瓷器装烧方法。这种方法可以提高产量、降低成本,并克服了器物容易变形的弱点。

(28) round ware(s)　圆器

(29) unglazed rim　芒口

(30) transparency　透明度

Exercises

Reading Comprehension Questions

Please answer the following questions according to the text.

(1) What are the characteristics of Xiuneisi wares according to *Ge Gu Yao Lun*?

(2) It is said in this text that among the Song Guan wares collected by the Palace Museum and the Taibei Palace Museum, daily use utensils are fewer while display ceramics account for the majority. Can you account for this phenomenon?

(3) What is the significance of the Chang brothers in the history of Longquan kiln porcelain, and how did they contribute to the development of this type of porcelain?

(4) What was the main type of porcelain produced by Jingdezhen kilns during the Song dynasty, and what characteristics made it renowned?

(5) How did the firing techniques and product characteristics of Jingdezhen qingbai porcelain evolve from the early Song dynasty to the mid and late Song dynasty?

Translation

A. Please translate the following paragraphs into Chinese, paying special attention to ceramic terminology, context and the social background involved.

(1) The Southern Song Guan ware not only inherited the simple yet dignified shape design and thick glaze texture of famous northern wares such as the Northern Song Bianjing Guan and Ru wares, but also assimilated the thin body combined with thick glaze, bright, clear glaze surface and exquisite shapes of famous southern wares such as Yue and Longquan wares. Cao Zhao of the Ming dynasty described Southern Song Guan ware in his *Ge Gu Yao Lun* as: "Products fired by Song dynasty's *Xiuneisi* (Guan kiln) were made from fine clay, had blue

color with pinkness of different shades, and featured crab claw patterns, purple rims and iron feet."

(2) It was because this porcelain type was created to imitate the color feature of bluish white jade that its glaze color was between blue and white. Hence, "*qingbai* (bluish white) porcelain" was used to refer to whitish blue ware or bluish white ware. It was called "*ying*（影）*qing* (shadow blue)" by the folks, and sometimes "*ying*（映）*qing* (reflected blue)," "*yin*（隐）*qing* (subtle blue)," and "zhao qing (covered blue)." The qingbai porcelain produced by Jingdezhen in the Song dynasty had a fine, white, delicate body, a bluish white glaze color resembling ice and jade, a color so bright and clear that it was able to show reflections. Mellow and crystalline, elegant and refined, such porcelain really deserved the praise "like beautiful jade" as recorded in literature.

B. Please translate the following paragraph into English.

宋代龙泉窑的兴起与越窑的衰败有着直接的关系。北宋前期，宁绍地区茶业兴盛，燃料匮乏，工匠雇值增加，窑业税收沉重等，使一代名窑在经历了千年的辉煌后在北宋中期走向衰落，由大规模的商品性生产转向就地销售的小规模生产。可能为了降低成本，产品制作也变得草率。在亚、非国家出土的越窑瓷器中，年代最晚的属于北宋中期，这说明越窑衰落后，其产品几乎不再外销。

Unit 6 Blue and White of the Yuan and Ming Dynasties

Blue and white wares were generally made by painting designs in cobalt oxide on the **unbaked body** of an object that was then covered with a clear glaze. Subsequently the piece was fired at a temperature in excess of about 1,250 ℃, which was sufficient to **vitrify** the clay and **fuse the glaze**. According to analyses of Yuan-dynasty blue and white wares from the Jingdezhen kiln center, the bodies were probably fabricated by adding an impure **kaolin** to a **kaolinite**-free porcelain stone.

The properties of these raw materials made the porcelain bodies easier to **throw** or **mold**—as well as to **finish** and fire—than the earlier Jingdezhen wares. Glazes were most likely made by adding limestone "glaze-ash" to the same porcelain stone. The technique of making blue and white wares became mature in the Yuan dynasty, but it reached the peak and popularized during the Ming dynasty.

Text

The Making of Blue and White in Late Yuan and Ming Dynasties

Jessica Harrison Hall

1 During the Tang dynasty, porcelains were chosen for use by the court from a variety of different kilns, including **Yue green wares** and **Xing white wares**. Imperial kilns, whose specific remit was to manufacture porcelains for the court, were set up in the Song dynasty. Following this earlier practice, the Yuan court set up the **Fouliang Porcelain Bureau** in 1278 to supervise the production of imperial ceramics. This office was located in the modern city of Jingdezhen, thousands of kilometers from the Yuan capital and an area renowned for its white rather than its green wares.

2 **Yuan dynasty**: Although cobalt was employed as a **colourant** and blue and white ceramics were made during the Tang and Song dynasties, it was in the late Yuan period between the 1320s and 1350s that large-scale high-fired manufacture of blue and white wares began at Jingdezhen. Blue pigment in the Yuan and early Ming dynasties (fourteenth to fifteenth centuries) used under court direction was imported from **Kashan** in Iran. It fires unevenly from black to pale blue, giving rise to the term '**heaped and piled**'. In the Yuan, official decorative motifs in any media were directly controlled by the **Hua Ju** (Design Office) of the **Jiangzuo Yuan** (Production Institute) in Dadu in the same way as the later Ming court relied on the Design Office under the **Board of Public Works**. Jingdezhen was not the only kiln site in China producing underglaze blue wares in the Yuan as was once generally accepted. Other kilns such as those at **Yuxi** in Yunnan also made blue and white wares using locally mined cobalt at that time, whose color was rather dull because of the high levels of manganese and iron. A major find of **Yunnanese** blue and white ware dating to the Yuan period was discovered in 1973 in the self-governing region of **Lufeng** to the west of Kunming in Yunnan. These wares were discovered in a series

of cremation tombs belonging to minority peoples.

3   Yongle porcelains are decorated with a particularly deep blue imported cobalt which fires unevenly to a dark blackish-blue in some parts and pale blue in others with "**silvery-black crystal-like spots**." This separation of colors, also observable in Yuan dynasty blue and white wares, is known as the "**heaped and piled**" **effect**. Locally mined cobalt could also be mixed with imported cobalt (probably from Iran) to alter the tone. The covering blue-tinged glaze is peppered with tiny pin pricks like the pores of skin. The porcelain bodies are finely **potted** and made of very pure **clay**. Some of the exquisitely painted designs derive from court painting styles for small **format album leaf** or fan paintings. **Zhe school artists** serving at the court may have influenced a stiffened style in painting at Jingdezhen which, although it lacks some of the vitality of the Yuan designs, is beautifully executed.

4   Xuande period: An innovation of the Xuande era was the successful firing of Chinese-mined cobalt called *potangqing*. This meant that the potters were no longer totally reliant on Near Eastern supplies. Uneven application and firing of the blue pigment with areas of pale blue, mid-blue and even black (the "heaped and piled" effect) are characteristic of the Xuande blue and white. Designs, which tend to cover the whole surface of the porcelain, are often slightly fuzzy or blurred as the cobalt proved difficult to control.

5   Zhengtong to Tianshun period: Underglaze blue wares of this period are decorated with local cobalt high in iron, which does not fire as attractively as the clearer blue imported cobalt. Vessels with figural designs are characterized by their "**wind swept**" **appearance**, which can be related to contemporary Ming fifteenth-century painting styles with designs outlined in dark blue and filled in with a paler blue **wash**. Many of the better-developed figural scenes which appear on *meiping* and large *guan* storage jars show designs which are based on popular legendary or immortal characters whose lives and exploits were described in books with woodblock-printed illustrations. These include the **San Guo Zhi Yan Yi** (Romance of the Three Kingdoms), Tang dynasty poems and tales of popular gods. More formal designs, such as dragons chasing flaming pearls among clouds, and stylized wave

and rock designs, continue traditions followed in the preceding Xuande era for imperial quality pieces. Calligraphically executed floral designs such as **lotus scroll** work or fruiting peach branches commonly appear on ***min yao*** (commercial kiln) products. Buddhist emblems and Sanskrit inscriptions appear on items such as dishes, jar-lets, stem cups and water bowls which were perhaps intended for ritual use. Designs for vessel forms and decoration employed in the civilian kilns appear to have enjoyed a long life, which makes their dating rather problematic.

6 Porcelain of the Chenghua period is mostly delicate, renowned for its pure glossy glaze and for the clarity of its underglaze blue decoration. Large areas are often left undecorated to show the purity of the porcelain and the clarity of the glaze. Elaborate border decorations are abandoned in the decoration of "palace bowls" whose designs are fuss-free, emphasizing the fineness and whiteness of the porcelain. Blue and white wares made in the Xuande era used imported cobalt which fired unevenly from pale blue to black, but from the late Chenghua era **Leping** cobalt, called *potangqing* or ***pingdeng***, mined in Jiangxi was used. This was much paler than imported cobalt, easier to control and gave a more even pigment and delicate lines. Underglaze blue motifs include phoenix, dragons, sea animals, the **Eight Treasures of Buddhism**, **double vajra**, children playing, the **Three Friends of Winter**, plants, insects, birds and Sanskrit inscriptions.

7 Hongzhi period: Among the ceramics ordered by the court, designs portraying dragons amid either lotus or clouds are the most common in underglaze blue. Traditional phoenix motifs were also popular. Flower and fruit ornament, together with Sanskrit characters from mantras, was often used too. The glaze is blue-tinged. Quality of porcelain body and of designs was not as high as in the preceding Chenghua era. Export orders account for a sizeable proportion of production in the Jingdezhen area during the Hongzhi period. For example, more than four hundred Chinese porcelains, mostly blue and white wares dating to the Hongzhi to Zhengde period, have been excavated from burial sites in **Calatagan**, **Batangas** Province, in the Philippines.

8 Blue and white imperial wares of the Zhengde era maybe divided into three main

types: those with Arabic and Persian inscriptions, those with dragon designs on a dense lotus ground or lotus designs and those archaistic items which copy fifteenth-century porcelains. There are also provincial *min yao* wares. Porcelains decorated with motifs specifically for the European market can be traced to the last years of the Zhengde emperor's reign. Generally, they are more **heavily potted** than porcelains of the Chenghua or Hongzhi reigns. These heavy forms are particularly noticeable among the porcelains with Arabic inscriptions. There is a greater experimentation with shapes reminiscent of the innovative years of the early fifteenth century. New forms of desk ware are introduced for Muslim eunuchs employed at court. Underglaze blue designs are achieved by outlining in blue and infilling with blue wash. According to the **Ruizhou Fuzhi** (Annals of Ruizhou), in the tenth year of Zhengde (1515) local cobalt from **Shanggao** was used. Later in the Zhengde reign imported cobalt was used once more. Zhengde period glaze is thick and blue-tinged.

9 Initially in Jiajing period blue pigment from Leping was used, but when the blue pigment was unavailable it was supplied from areas of **Ruizhou**. Imported cobalt was superior to the local product and was, by contrast to the pale local blue, quite vivid. When ground for a long time it could appear almost violet in color. Supplies of imported cobalt were mixed with local pigment and graded according to the proportions of local cobalt. The ideal mix contained just one tenth of local cobalt and the poorest grade was nine-tenths local product. Sixteenth-century palace order registers of the Jiajing reign after 1529 exist detailing both the volume of porcelains and the shapes and decorative designs to be used. For example, in the year 1544, 1,340 **services** were ordered comprising a total of about 36,000 pieces, including fruit dishes, vegetable dishes, rice bowls, teacups, wine cups, sauce dishes and **leys or slops jars**. Blue and white wares made for export are understandably made of less high-quality materials. Such porcelains have been discovered in the Philippines, Indonesia, Egypt, East Africa and the Near East. Overseas trade development, particularly in blue and white wares, was a fundamental achievement of the Jiajing era.

10 In Longqing period, Underglaze blue wares are very similar in style to those of

the Jiajing era but as the reign is so much shorter (six as opposed to forty-five years) far fewer of these were made. Cobalt was imported and of a very dark ink blue. Boxes were produced in a variety of forms: square, rectangular, round, **seven-lobed** and ingot-shaped. Other shapes include large vats for fish rearing, jars, incense burners, flower shaped basins, bowls and dishes in a variety of forms. Most of the designs are traditional dragon, phoenix and flower motifs. This repertoire is also reflected in the lacquer wares of the period. Figures are rather elegant and elongated.

11 A shard dating to Wanli period was excavated in Ming **remains** at **Zhushan**, Jingdezhen, showing a carp leaping in underglaze red from waves painted in underglaze blue.

12 Early in the reign imported ***huiging*** (Mohammedan blue cobalt) was used, but later, because supplies dwindled, local Chinese pigment from Zhejiang, Jiangxi, Guangxi, Anhui and Fujian was employed. According to the ***MingShilu***, in 1596 blue pigment was ordered to be sent as tribute from the vice governor of **Gansu**, but in 1605 Zhejiang pigment was used. Local cobalt was purified using magnets to extract the iron content. The usage of local and imported cobalt is further complicated as imported blue and local cobalt were mixed together in different ratios, depending on their quality. Clay used in the early Wanli reign from **Macang** was exhausted in 1584 after which poorer clay had to be brought from the **Kaihua** mountains.

Text Related Information

This text is adapted from the book *Catalogue of Late Yuan and Ming Ceramics in the British Museum* (2001), written by Jessica Harrison Hall, published by the British Museum Press, London.

Selected Plates Related to the Text

Plate 28 *Meiping* with cover with underglaze blue decoration. Height with cover: 44.5 cm, width: c. 22 cm, Yuan dynasty.

Plate 29 Bottle with underglaze blue decoration and cut-down neck. Height: 27 cm, Ming dynasty.

Unit 6 Blue and White of the Yuan and Ming Dynasties

Plate 30   Octagonal brush and ink-stick stand decorated in underglaze blue. Height: 6.2 cm, diameter: 12 cm, Ming dynasty, Jiajing mark and period.

Plate 31   Dish with underglaze blue decoration. Height: 3.7 cm, diameter: 28.5 cm, Ming dynasty.

Plate 32   Large bowl with underglaze blue decoration. Height: 12 cm, diameter: 30.3 cm, Ming dynasty, Yongle period.

Aids to Comprehension

Notes

（1）**Fouliang Porcelain Bureau**　浮梁瓷局,是元朝在景德镇设置的唯一一所为皇室服务的瓷局。浮梁瓷局秩正九品,至元十五年(公元1278年)立,掌烧造瓷器,并漆造马尾、棕藤笠帽等事。在瓷局的掌管下,枢府釉、青花、釉里红、蓝釉、蓝地白花、孔雀蓝釉等瓷器得以烧造。

（2）**Yuan dynasty**　元朝(1271—1368年),是中国历史上的朝代。铁木真于1206年建国;1271年忽必烈定国号为元,1279年灭南宋。元朝是首个由少数民族建立的大一统王朝,传五世十一帝。

（3）**Kashan**　卡尚,伊朗中部的一个城市,位于扎格罗斯山东麓、面积50平方公里的卡尚绿洲中,海拔994米,人口11万。

（4）**Hua Ju**　画局,元代宫廷专门负责和管理绘画方面事物的机构。

（5）**the Jiangzuo Yuan**　将作院,元官署名,主官秩正二品。其所辖各种工艺品的作场非常繁多,"掌制造金玉珠翠犀象宝贝冠佩器皿,织造刺绣缎匹纱罗,异样百色造作"。至元三十年(1293年)始置,有院使、经历、都事等官。所属有诸路金玉人匠总管府、异样局总管府、大都等路民匠总管府等。各总管府又分辖诸局、所、司、库等。

（6）**Board of Public Works**　工部,中国封建时代中央官署名,为掌管营造工程事项的机关,六部之一,长官为工部尚书,曾称冬官、大司空等。

（7）**Yuxi**　玉溪,在今云南。玉溪窑是明代景德镇窑以外生产青花瓷的重要窑场,始烧于宋元,而止于明,共发现三处窑址,均烧青釉和青花瓷器。青釉有印花、划花及无纹饰三种,印花多阳纹花卉,划花为云纹与水波纹。青花瓷器釉色与青瓷相同,纹饰有鱼藻、折枝花卉及四佛杵等纹。器型有大碗、大盘等,还有玉壶春瓶、罐等。

（8）**Yunnanese**　云南产的,这里指云南产的青花瓷。

（9）**Lufeng**　禄丰,属楚雄彝族自治州。

（10）**San Guo Zhi Yan Yi**　《三国演义》,又名《三国志演义》《三国志通俗演义》,是元末明初小说家罗贯中创作的长篇章回体历史演义小说,与《西游记》《水浒传》《红楼梦》并称为中国古典四大名著。

(11) min yao　民窑,与官窑相对而存在。官窑主要是为皇家和贵族生产瓷器,其生产数量较少,质量标准严格,追求精良的工艺和华丽的装饰,以满足上层阶级的需求。民窑则主要面向民间市场,生产规模较大,门槛较低,追求实用性和多样性,以满足普通百姓的需求。

(12) Leping　乐平,隶属于江西省景德镇市,位于江西省东北部,地处鄱阳湖盆地边缘与赣北丘陵交界处。

(13) Eight Treasures of Buddhism　佛家八宝,又称八瑞相、八吉祥,包括宝瓶、宝盖、双鱼、莲花、右旋螺、吉祥结、尊胜幢、法轮,均为表示吉庆祥瑞之物,象征吉祥、幸福、圆满。佛家八宝是陶瓷装饰图案中常见的形式。

(14) double vajra　十字金刚杵,寓意坚固、锋利、智慧、去除烦恼、消除恶魔。所谓十字金刚杵,实际上是由两个金刚杵相交叉构成的,因形似"十"字而得名。十字金刚杵的四个杵头分别为白(东)、黄(南)、红(西)、绿(北)四色。十字金刚杵的中心,即两杵的交汇点,呈圆形,为蓝色。

(15) Three Friends of Winter　岁寒三友,陶瓷器物装饰图案中常见的一种图案形式。岁寒三友指松、竹、梅三种植物。松、竹经冬不凋,梅花耐寒开放。

(16) Calatagan, Batangas　菲律宾八打雁省卡拉塔甘市。八打雁是菲律宾的一个省份,位于吕宋岛的西南部,距离菲律宾首都马尼拉大约90千米,交通便利。卡拉塔甘市是八打雁省的一个直辖市。

(17) Ruizhou Fuzhi　《瑞州府志》,共十四卷,明熊相、邝璠纂修,现有明正德刻本收藏于宁波市天一阁博物院。

(18) Shanggao　上高县,隶属江西省宜春市,位于江西省西北部,地处赣江支流锦江中游,东与高安市为邻,西与万载县接壤,南和新余市渝水区、分宜县交界,北与宜丰县相连。

(19) Ruizhou　瑞州,现江西省高安市。

(20) Zhushan　珠山,指景德镇珠山区,是景德镇御窑厂遗址所在地。

(21) MingShilu　《明实录》,是明朝历朝官修的编年体史书。自明太祖朱元璋到明熹宗朱由校共15朝13部,2900多卷,1600多万字,具有重要的史学价值,是研究明朝历史的基础史籍之一。

(22) Gansu   甘肃省

(23) Macang   麻仓,景德镇附近的麻仓山出产的一种高岭土。麻仓土质量优良,是制造瓷器的主要原料之一。

(24) Kaihua   开化山。这里指浙江省开化县窑址所在区域。

Ceramic Terminology

(1) blue and white   青花瓷,又称白地青花瓷,简称青花,是中国瓷器的主要品种之一,属釉下彩瓷。青花瓷是以含氧化钴的钴矿为原料,在陶瓷坯体上描绘纹饰,再罩上一层透明釉,经高温还原焰一次烧成的。钴料烧成后呈蓝色,具有着色力强、发色鲜艳、烧成率高、呈色稳定的特点。原始青花瓷于唐宋已见端倪,成熟的青花瓷则出现在元代景德镇的湖田窑。明代青花成为瓷器的主流,宣德时发展到了顶峰。明清时期还创烧了青花五彩、孔雀绿釉青花、豆青釉青花、青花红彩、黄地青花、哥釉青花等衍生品种。

(2) the unbaked body   未烧制的坯体,指青花瓷在着色上釉前的白色陶坯。瓷器在上釉之前一律称为陶坯或素坯,以高岭土制成,有很浅淡的光泽。这是进行彩绘和上釉之前的原始材料。

(3) vitrify   玻璃化,这里指烧结。

(4) fuse the glaze   熔融釉料

(5) kaolin   高岭土,是高岭石经过加工处理得到的微粒状黏土,常呈白色或微黄色,比高岭石更细腻、柔软和易碎。

(6) kaolinite   高岭石,一种天然的铝硅酸盐矿物,通常呈白色或淡黄色。主要化学成分为氧化铝、硅酸和水,其中氧化铝含量达到30%以上。高岭石的晶体较大,通常呈壳状或柱状,不易碎裂,比较硬。

(7) throw   拉坯成型,陶瓷术语。拉坯成型是利用拉坯机旋转的力量,配合双手的动作,将转盘上的泥团拉成各种形状的成型方法,也叫轮制法。它是利用拉坯机快速旋转所产生的离心力,结合泥巴的特性与手和拉坯机之间的动力规律,用手控制和挤压泥团,制成空腔薄壁的圆体器型。拉坯是陶艺制作中较常用的一种方法。

(8) mold   模制成型法,陶瓷术语,模制法是以模具为依托的成型方法。现代陶艺使用的模具一般可分为印坯模和注浆模两种。

（9）finish　修坯；利坯。制作陶瓷的坯体,由于表面不太光滑,有的还有模缝迹或流浆等情况,因此需要进一步加工修平,这道工序称为修坯。修坯在旋车上操作完成,旋车中心立有木桩,桩顶端呈圆形。修坯时将坯放于桩上,旋转后用车刀旋之,使其里光外平,最后把底部多余的部分修掉并挖足。

（10）Yue green wares　越窑青瓷。越窑是中国古代最著名的青瓷窑系,其主要产地是唐代明州慈溪县(今浙江省宁波市慈溪市)上林湖一带,因五代时划归越州而得名"越窑"。越窑持续烧制了1000多年,于北宋末、南宋初停烧,是中国持续时间最长、影响范围最广的窑系。

（11）Xing white wares　邢窑白瓷。邢窑是中国白瓷的发祥地,中国古代最早的官窑之一。其产地河北内丘在唐代属于邢州,故史称邢窑。邢窑是中国最早的白瓷窑址,邢窑白瓷的发明与制作,打破了自商代以来,青瓷一统天下的局面,形成了南方以浙江慈溪越窑为代表的青瓷和北方以河北内丘邢窑为代表的白瓷并驾齐驱、平分秋色的"南青北白"的格局。

（12）colourant　着色剂。这里指绘制青花瓷的青花料,即氧化钴,也就是钴土矿。青花瓷的釉下花纹是以氧化钴为着色剂来描绘装饰的,大体分为含锰量高、含铁量低的国产青花料和含锰量低、含铁量高的进口青花料。国产青花料主要有珠明料、浙料、石子青、平等青(陂塘青)等。国产青花料资源较为丰富,主要产地为江西赣州、上高、乐平、上饶等,浙江、广东、云南、福建、广西等地也有丰富的钴土矿。进口钴料主要是来自古代波斯的苏麻离青(也叫苏料)和来自西域的回青料。

（13）heaped and piled　结晶斑。元青花瓷所用的颜料含有杂质,因此在绘画和烧制过程中颜料色粒会向某处聚集而形成人们常说的积青处(颜色深重发黑,也有的称凝聚斑)。

（14）silvery-black crystal-like spots　银色透明的结晶

（15）"heaped and piled" effect　堆积效果,一种青花呈色不均匀的现象,即某些地方呈色较深,有些地方呈色较浅。

（16）potted　胎体形状工整,制作精良

（17）clay　黏土

（18）format album leaf　小册页

(19) Zhe school artist　浙派艺术家

(20) Potangqing　陂塘青,亦称"平等青",国产青花料中的一种。明代成化、弘治、正德早期景德镇窑青花多用此料。平等青含铁量较少,烧成后,色泽淡雅、清丽而明澈,晕散不严重,呈色淡雅青亮,与明初永乐、宣德呈色浓重青翠的苏麻离青截然不同。

(21) "wind swept" appearance　微风吹拂风格,一种绘画风格。

(22) wash　晕染,这里指填涂薄层青料。

(23) lotus scroll　缠枝莲图案

(24) pingdeng　平等青,青料的一种,即陂塘青。

(25) heavily potted　胎体厚重

(26) services　指餐具套件

(27) leys or slops jar(s)　渣斗,又名爹斗或唾壶,主要用于盛装唾吐物,如置于餐桌上,专用于盛载骨头、鱼刺等食物渣滓。小型者亦用于盛载茶渣,因此也列于茶具之中。

(28) seven-lobed　七瓣形,指器物形状呈花瓣形。

(29) remains　遗存,文物学术语,包括遗迹、遗物。遗迹指人工建造的各种工程和遗留下来的各种痕迹,如聚落(村落、城市)、建筑、墓葬、道路等;遗物指人类制作和使用的各类物品,如工具、武器、日用器具、装饰品等。

(30) Huiqing　回回青,也称回青,是一种来自阿拉伯地区的进口青花原料,有时也被称为苏麻离青或者苏泥勃青。

Exercises

Reading Comprehension Questions

Please answer the following questions according to the text.
(1) What is the making process of blue and white?
(2) What are the features of Yongle blue and white?
(3) What are the characteristics of "wind swept" appearance?
(4) What are the features of Chenghua blue and white?
(5) What are the three main types of blue and white in Zhengde era?

Translation

A. Please translate the following paragraphs into Chinese, paying special attention to ceramic terminology, context and the social background involved.

(1) Blue and white wares were generally made by painting designs in cobalt oxide on the unbaked body of an object that was then covered with a clear glaze. Subsequently the piece was fired at a temperature in excess of about 1250 degrees, which was sufficient to vitrify the clay and fuse the glaze. According to analyses of Yuan-dynasty blue and white wares from the Jingdezhen kiln centre, the bodies were probably fabricated by adding an impure kaolin to a kaolinite-free porcelain stone. The properties of these raw materials made the porcelain bodies easier to throw or mold—as well as to finish and fire—than the earlier Jingdezhen wares. Glazes were most likely made by adding limestone "glaze-ash" to the same porcelain stone.

(2) Blue and white imperial wares of the Zhengde era maybe divided into three main types: those with Arabic and Persian inscriptions, those with dragon designs on a dense lotus ground or lotus designs and those archaistic items which copy fifteenth-century porcelains. There are also provincial *min yao* wares.

Porcelains decorated with motifs specifically for the European market can be traced to the last years of the Zhengde emperor's reign. Generally, they are more heavily potted than porcelains of the Chenghua or Hongzhi reigns. These heavy forms are particularly noticeable among the porcelains with Arabic inscriptions. There is a greater experimentation with shapes reminiscent of the innovative years of the early fifteenth century. New forms of desk ware are introduced for Muslim eunuchs employed at court. Underglaze blue designs are achieved by outlining in blue and infilling with blue wash.

B. Please translate the following paragraph into English.

说到"瓷都"景德镇对中国陶瓷的贡献，首先得从元代景德镇烧制的青花说起。元代青花瓷产地，还有浙江江山、云南建水及玉溪等地。但无论产量、质量还是艺术价值，它们都无法与景德镇烧制的青花瓷媲美。青花瓷是一种以天然钴土矿为呈色剂，在白釉坯体上用毛笔描绘图案花纹，罩透明釉后，再入窑以高温一次烧成的釉下彩瓷。它创烧于唐代河南巩县窑，到元代臻于成熟，色彩清雅，如中国传统水墨画般极具风韵。

Unit 7　Polychrome Glazed Wares in the Qing Dynasty

Painted polychrome ceramic wares refer to all types of ceramic wares with various painted decoration in polychrome colors. In the category of pottery wares, there are color-glazed pottery and painted pottery wares. In the category of porcelain wares, there are wares in underglaze colors, overglaze colors, combination of both underglaze and overglaze colors, wares in other color glazes and enamels, plain three-colors, etc. .Underglaze colored wares include wares in underglaze brown, underglaze brown with underglaze green, underglaze black, underglaze blue, underglaze red, combination of underglaze blue and red, combination of cobalt blue, copper-red, and peagreen (wares in three underglaze colors), underglaze five colors, etc. Overglaze colored wares include wares painted in five-color enamels, wares with porcelain paste and painted in opaque enamels, wares in *yangcai* enamels, wares in *fencai* enamels, etc. The combination of both underglaze and overglaze colors include wares in underglaze blue and *wucai* enamels, *doucai* wares, wares in underglaze blue and *fencai* enamels, etc. Wares in other color glazes and enamels refer to wares with single color-glazed ground and painted with decorations in only one single color glaze or enamel. Wares in **plain three-colors** are those decorated with three or more low-temperature fired glazes or enamels but without the use of red color.

Text
The Polychrome Glazed Wares in the Qing Dynasty
Suzanne G. Valenstein

1 The majority of the finest Qing-dynasty wares were produced during a comparatively short period, from 1683 until about 1756, when the imperial factories at Jingdezhen were under the direction of several capable supervisors; the most famous were **Zang Yingxuan**, **Nian Xiyao** and **Tang Ying**. During their incumbencies, the potting methods that were built on the accumulated experience of the past were developed and refined to produce porcelain that is the apogee of perfection. Old decorative techniques were enlarged and improved upon, and new ones were added to the traditional vocabulary. Frequently, several of these ornamental devices were combined on one piece, thus producing a kaleidoscope of monochromatic and polychromatic effects.

2 Enumerating all the types of porcelains produced in the Qing period is a formidable task, and, while most of them are discussed here, not every one can be mentioned.

3 **Famille verte enamels**. The possibilities of painting porcelains with colored enamels, which were so successfully developed during the Ming dynasty, were exploited to the fullest in the Qing period. Essentially the same low-fired enamels were now used with such authority that Qing polychrome-decorated porcelains outshine all others. The stellar polychrome decoration of the Kangxi period, the famille verte palette of enamels, takes its name from the several distinctive shades of green that are almost invariably present in the color scheme. Famille verte enamels are brightly colored and translucent; they have been applied rather thickly over the darker outlines and details. In addition to the various greens, the famille verte colors include yellow, aubergine, coral toned iron-red (rather flat and almost opaque); white (achieved by allowing the pure body to show through a clear enamel); and black (a composite color made of matte, brownish black pigment covered with green, aubergine, or clear enamel). The blue enamel in this assortment of colors is different

from the Ming-dynasty turquoise-tinted blue enamel; it is more violet or royal blue in tone. Like their Ming antecedents, these translucent famille verte enamels—appropriately named *yingcai* ("hard colors") by the Chinese—did not permit much gradation in color, and the effects of shading had to be relegated to finely penciled lines in the preliminary drawing.

4 Famille verte enamels were painted directly onto unglazed, prefired bodies, known as **enameling** on the biscuit, or they were painted over high-fired **clear glazes**. In both instances, the decorated porcelains, which had already been high fired to maturity, were given a second firing. This firing was only to the low temperatures of an enameling kiln, known in the West as a **muffle kiln**, to fuse the enamels.

5 Famille verte enamels painted on the **biscuit**. The decoration of porcelains painted with famille verte enamels on the biscuit is often set against a dominant, **solid-color ground** of black, yellow, green or occasionally, aubergine; the first two groups are frequently placed in the subcategories of famille noire and famille jaune. At their peak, these vessels are truly gorgeous, combining eloquent brushwork with well-mated colors and designs that are invariably right for the shape they grace. Three vases, representing the famille noire group, the famille jaune class, and the rare family with green grounds, are excellent examples of their respective types. They are the sort of porcelains that were greatly treasured by past generations in both this country and Europe.

6 Enamel painting on the biscuit was particularly well suited to the decoration of porcelain figures because a thick, preliminary glaze under the enamels would tend to fill in and blunt the sharp modeling of the features and the garments. Two superb figures are masterful examples of **modeling** and decoration; they are at the summit of Kangxi porcelain figures. The iron-red on one of these figures has been painted on a pad of clear glaze that was applied just in that area to support the red, a technique generally employed in on-the-biscuit enameling to ensure a richer red color.

7 A certain amount of the **openwork** that was popular in the late Ming period is found again in famille verte on-the-biscuit wares, particularly in such objects as

writing cases, brush holders, and lanterns.

8   Famille verte enamels painted over the glaze. When used over the glaze, the famille verte enamels present a somewhat different appearance. Supported by the lustrous glaze, they stand radiant and clear against the white ground that forms an integral part of the total **composition**. In addition to being used with the usual overglaze blue enamel of the palette, overglaze famille verte enamels are sometimes found in conjunction with underglaze cobalt-blue painting, and occasionally both underglaze and overglaze blue can be seen on the same object. Touches of gilt were often added to this group, providing an especially lively accent.

9   The designer's imagination has seldom been more fruitful or wider in scope than on these sumptuous porcelains; they offer a galaxy of **motifs** handled in an almost infinite variety of ways. Their decoration ranges from relatively few figures discreetly placed on broad expanses of white, in a style most likely to have appealed to the court, to extremely complex designs featuring **reserve panels** of small scenes set against the richly brocaded grounds that virtually cover the vessel's surface.

10   **Sang de boeuf glazes**. The first is the intense, brilliant, red glaze known as **Langyao**, sang de boeuf, or "ox blood." In its finer examples, this spectacular glaze gives the impression that one is gazing through a limpid surface layer, which is slightly crazed and strewn with countless fine bubbles, to the color that lies underneath. Sang de boeuf color generally starts in a greenish gray tone at the top of the vessel; as it descends, it quickly turns red in changing shades that range from light red with tints of green to deep crimson, with an occasional overtone of dark reddish brown. A remarkable control of the thick glaze, which is checked in an even line where it stops above the foot, is considered to be a hallmark of genuine Kangxi Sang de boeuf porcelains. The small vase is an especially fine example of this type.

11   **Peachbloom glazes**. In the second copper-red glaze, the famous Kangxi peachbloom glaze, the effect is quite subdued. This soft, velvety glaze varies in color from piece to piece, but it essentially is pale pinkish red, often shading to darker values. It sometimes is plain, frequently is mottled, and in a particularly appealing version, it shows tender flushes of mossgreen. The finest peachbloom

wares constitute an elite series—in all likelihood consisting of no more than eight specific shapes—of small and elegant vessels that were intended to be used at the scholar's writing table. The refinement in the **potting**, shapes, and glaze in this group indicates that it probably dates to the final portion of the Kangxi reign. Not many collections can boast of having more than a few classic peachbloom examples, and the Metropolitan is fortunate to own seven of the eight basic forms and what most likely is a variation of the eighth.

12　**Clair de lune glazes**. The Kangxi period yielded a broad spectrum of high-fired blue glazes, which derived their color from cobalt oxide. They range from a fairly purplish midnight color through many lovely intermediate shades to the fragile tint known as clair de lune. Clair de lune-glazed porcelains frequently have the same shapes as the classic peachbloom vessels and, like the latter, show the daintiness and fineness of potting that is associated with the end of the Kangxi period. Three objects for the writing table—a water coupe, vase and brush washer—are sheathed with a gentle, even glaze that shows just a breath of color. They illustrate why clair de lune is among the most treasured of the Qing-dynasty glazes.

13　**"Powder-blue" glazes**. "Powder-blue" glazes are somewhat different from other Qing blue glazes in that the cobalt coloring matter was not mixed with the glaze. Instead, it was blown dry onto the raw body of the vessel by using a piece of gauze stretched over the end of a bamboo tube. This was then covered with a clear glaze and fired. The result is a somewhat frothy effect, at its best a rich **lapis lazuli** tone, which might be further embellished with etched designs or, more often, gilding. Such gilding is seen on the imposing vase, where the gilt is unusually well preserved. One of the poems on this vase is dated, "On a spring day of the year **jichou**" (1709). "Powder-blue" has frequently been used as a ground against which reserve white panels of various shapes, which contain sundry underglaze blue or overglaze polychrome enamel designs, are set.

14　**Famille rose**. The signal accomplishment of the Yongzheng period was porcelain decorated in the famille rose palette of enamels. These opaque and semiopaque enamels differ from the earlier overglaze polychrome enamels in two respects. First is the addition of a rose pink, from which this type of enameling takes its name. This pink, which was derived from colloidal gold, was used in a wide spectrum of tones from the palest blush of pink to deep ruby. Second—and perhaps even more important—a **lead-arsenic**, opaque white pigment was mixed with the colors to modify them, enabling the painter to achieve a range of color values for the first time. These new graduated tones allowed the artist to reproduce subtleties of shades and to model his drawing as the artist who paints with oils does. In addition, a variety of mixed tints was produced by combining colors, usually with the addition of the **opaque white**. Iron-red was occasionally added to the gentle hues of the famille rose palette, and this combination of harsh and soft colors can be surprisingly effective. The Chinese call the low-fired famille rose palette *fencai* ("pale colors"), *ruancai* ("soft colors"), *yangcai* ("foreign colors"), or *falangcai* ("enamel color"). Plate 33 and 34 are both classics of Famille rose wares.

15 Apparently the technique of painting porcelain in these famille rose enamels was known in Europe before it was known in China. For example, some Viennese porcelains of the **Du Paquier period** that might be dated to about 1725 illustrate a rather skillful use of shaded overglaze enamels on porcelain bodies.

16 In China, the famille rose palette could have been adopted from the European painted enamels on gold and copper that were introduced to China by Jesuit missionaries; however, it could also have been adopted from the decoration on South German tin-glazed earthenwares brought by the Dutch.

17 There is some difference of opinion as to precisely when the Chinese were able to perfect the use of these enamels on porcelain. Although rudimentary famille rose enamels can be found on some **armorial porcelains** dating to the early 1720s, most likely the technique was not fully developed much before 1730. Once established, however, the famille rose style of painting became the dominant factor in the Chinese decorative vocabulary.

18 The finest famille rose porcelains of the Yongzheng period, which were probably made in imperial kilns at Jingdezhen for the exclusive use of the emperor and his court, are the epitome of delicacy and restraint in painting. The enameling on these

"court-taste," or "Chinese-taste," wares is exquisite; the designs generally are sparse, allowing the beautiful white body to play an important part in the composition. A particular whimsy sometimes found on this type of porcelain may be seen on the impeccable bowl starting on the outside, the picture continues over the rim and is completed on the interior surface.

Text Related Information

This text is adapted from chapter twelve of the book *A Handbook of Chinese Ceramics* (1989), written by Suzanne G. Valenstein, published by the Metropolitan Museum of Art, New York.

Selected Plates Related to the Text

Plate 33   Figure, possibly the God of Wealth. It is painted in famille verte enamels on the biscuit and on the glaze. Height: 60.6 cm. Qing dynaty, late 17th-early 18th century, Kangxi period.

Plate 34　Vase on the left: Porcelain with Sang de boeuf glaze. Height: 20 cm. Qing dynasty, late 17th-early 18th century, Kangxi period. Vase in the middle: Porcelain painted in underglaze red. Height: 19.7 cm. Qing dynasty, Yongzheng mark and period. Vase on the right: Porcelain with peachbloom glaze. Height: 20.3 cm. Qing dynasty, Kangxi mark, late in the period.

Plate 35　Vase. Porcelain with "power-blue" glaze, painted in gilt. Height: 44.5 cm. Qing dynasty, early 18th century (poem dated in accordance with 1709).

Unit 7 Polychrome Glazed Wares in the Qing Dynasty

Plate 36 Dish. Porcelain painted in overglaze famille rose enamels, crimson pink glaze on the reverse. Qing dynasty.

Plate 37 Bowl. Porcelain painted in overglaze famille rose enamels. Qing dynasty, Yongzheng mark and period.

Aids to Comprehension

Notes

(1) **Zang Yingxuan** 臧应选,清代三大督陶官之一。他在景德镇御窑厂工作期间,不仅提升了陶瓷的生产技术,还推动了陶瓷设计的创新,使清代的陶瓷产品更加精美和多样化。

(2) **Nian Xiyao** 年希尧,清代三大督陶官之一。作为督陶官,年希尧注重陶瓷的质量和工艺,他的管理使得御窑瓷的产量和质量都有了显著的提升。

(3) **Tang Ying** 唐英,清代三大督陶官之一。唐英在景德镇御窑厂任职期间,以卓越的管理才能和创新能力,被誉为"唐窑"的创始人。他在陶瓷的制作技术、设计以及管理方面都有重大的贡献,使清代的陶瓷艺术达到了一个新的高度。

(4) **Jichou** 农历己丑年。这里指康熙四十八年,即公元 1709 年。

(5) **Du Paquier** 迪帕基耶瓷器。维也纳的迪帕基耶瓷器厂是继德国梅森瓷器厂之后第二个成功掌握瓷器"秘密"的欧洲工厂。

Ceramic Terminology

(1) **polychrome glazed ware(s)** 颜色釉陶瓷或多彩釉陶瓷。彩釉可以分为釉下彩和釉上彩两大类。釉下彩,也称"窑彩",是在已成型、晾干的素坯(半成品)上绘制各种纹饰,然后罩以白色透明釉或其他浅色釉一次烧成的。烧成后的图案被一层透明的釉覆盖,表面光亮、柔和、平滑,显得晶莹透亮。釉上彩(如五彩、粉彩等),是在已烧成瓷的釉面上描绘纹样、填彩,再入窑以低温烧成的一种陶瓷上色技术,烧成温度为 700 ℃—800 ℃。彩瓷的发展经历了从釉下彩到釉上彩的过程,尤其是在明清时期,景德镇的彩瓷品种达到了数十种甚至上百种,显示了彩瓷艺术的丰富多样性和深厚的历史底蕴。

(2) **yangcai, fencai, wucai, doucai** 洋彩、粉彩、五彩、斗彩。洋彩也称为新彩,

是釉上彩的一种，包括贴花、绘画、刷花、喷花、印花、薄膜移花、描金加彩、套色印金、腐蚀金彩和各色电光彩等。粉彩又称软彩，一般情况下是在高温烧制的白瓷上勾勒图案，使用含砷的玻璃白打底，再根据深浅浓淡的不同需要，用干净的笔轻轻地把颜料施于玻璃白之上，使花瓣和人物衣服有浓淡明暗的层次感。五彩是釉上彩的一种，具体做法是在烧好的素瓷釉面上进行彩绘，再入窑经 600 ℃—900 ℃ 温度烧成的一种瓷器。五彩指分布在瓷器釉面上多种色彩，而五彩瓷并不一定指瓷器釉面上只有五种颜色。多于或少于五种颜色的陶瓷，在习惯上也同样称为五彩瓷。五彩瓷在明清两代发明和发展，其配方经过不断创新，才出现以红、黄、绿、蓝、黑、紫等为主的彩瓷。斗彩又称逗彩，创烧于明朝宣德年间，明成化时期的斗彩最受推崇，是釉下彩（青花）与釉上彩相结合的一个装饰品种。斗彩是预先在高温（1300 ℃）下烧成的釉下青花瓷器上，用矿物颜料进行二次施彩，填补青花图案留下的空白和涂染青花轮廓线内的空间，然后再次入窑经过低温（800 ℃）烧成的。

（3）plain three-colors　素三彩，瓷器釉彩名，在未上釉的素胎上，施以绿、黄、茄紫三色而烧成。素三彩始于明正德年间，清康熙时继续烧制。其制作方法是，在高温烧成的素胎上用彩釉填在已刻划好的纹样内，再经低温烧成。

（4）famille verte enamels　硬彩，又名古彩。为区别于清代粉彩，一般称明代五彩和清代康熙五彩为古彩或硬彩。五彩，是将红、黄、绿、蓝、紫等各种带玻璃质的彩料，按图案纹饰的需要施于釉上，在瓷胎上用生料、矾红勾线，用单线平涂，再入窑二次烧成的一种古彩。它红绿分明，层次较少，彩色鲜明、透彻，故又称硬彩。

（5）enameling　彩饰；上光。釉上彩装饰中的彩饰是指在已经烧成的瓷器釉面上，使用各种彩料绘制各种纹饰，然后二次入窑并在低温下固化彩料的过程。这是陶瓷的主要装饰技法之一，通常包括彩绘瓷、彩饰瓷、青花加彩瓷、五彩瓷、粉彩瓷、色地描金瓷及珐琅彩等。釉上彩的历史可以追溯到公元575年，当时这种技术被称为"炉彩"或"釉上加彩"。彩饰不仅增加了陶瓷的美观性，而且为陶瓷艺术增添了丰富的色彩和图案。

（6）clear glaze　透明釉，也称为清釉，主要采用天然原料（如长石、石英、石灰石和高岭土等）配制而成。清釉不起盖底作用，坯体本身的颜色能通过釉层反映出来,这种釉料的特点是成本低,工艺简单,制造容易,烧成范围宽,性能稳定。涂在物体表面的清釉,在干燥后形成光滑的薄膜,显出物体表面原有的纹理。

(7) muffle kiln　马弗窑,又称隔焰窑,是用隔焰板将热源(燃烧产物或电热元件)与坯体隔开,热源将隔焰板加热,借隔焰板的辐射作用将热传给坯体,使坯体烧成的一类窑。

(8) biscuit: unglazed pottery that has been fired　素胎;未施釉的烧制过的胎体

(9) solid-color ground　单色地

(10) modeling　造型

(11) openwork　镂空雕刻

(12) composition　构图

(13) motif　图案

(14) reserve　(陶瓷装饰或纺织品印染中保留原色的)防染本色区

(15) panel　开光,是一种在瓷器上通过几何形状框出特定的区域,并在其中填充不同图案的装饰技术。这种手法通过轮廓内外的形象、色彩、肌理等对比,创造出强烈的视觉效果,使得装饰主题更加突出,具有艺术感召力。

(16) sang de boeuf glaze　牛血红釉(法语)

(17) Langyao　郎窑。郎窑红是中国名贵铜红釉中色彩最鲜艳的一种,其特点是色彩绚丽,红艳鲜明,具有一种强烈的玻璃光泽。由于釉汁厚,且在高温下流淌,口沿处往往露出白胎,呈现出圈状白线,俗称"灯草边"。郎窑红底部边缘釉汁流垂凝聚,近乎黑红色。为了流釉不过底足,工匠用刮刀在圈足外侧刮出一个二层台,阻挡釉流淌下来。这是郎窑红瓷器制作过程中的一个独特技法,因此有"脱口垂足郎不流"之称。

(18) peachbloom glaze　豇豆红釉、桃色釉,亦称"美人醉",清康熙时铜红釉的名贵品种之一。烧制时先在坯上施一层底釉,然后吹上一层颜色釉,再盖上一层面釉,入窑以高温还原焰烧成,呈色变化较大。粉红色中略带灰色的称"豇豆红釉",灰而色暗的称"乳鼠皮釉",粉红中有绿点的称"苔点绿釉",带红块的称"孩儿面釉"。

(19) potting　制作

(20) clair de lune glaze　月白釉。宋代,月白釉在瓷器世界大放异彩,尤其是钧窑月白釉更是有名。钧窑的基本釉色是各种浓淡不一的蓝色乳光釉,较深的称天蓝,较淡的称天青,比天青更淡的称月白。总体而言,月白就是带乳白或乳光感的浅淡蓝色。

(21) "powder-blue" glaze 粉末绀青,也叫吹青,瓷器釉色名,清代康熙朝景德镇仿烧明宣德瓷的一个蓝釉品种。

(22) lapis lazuli 青金石。一种半宝石,通常呈深蓝色,主要成分是复杂的硅酸盐,常常带有黄铁矿斑点。

(23) famille rose 粉彩、软彩。粉彩瓷是珐琅彩之外,清朝宫廷创烧的一种彩瓷。通常是在烧好的胎釉上施含砷的粉底,涂上颜料后用笔洗开。由于砷的乳浊作用,颜料会产生粉化效果。

(24) falangcai 珐琅彩。珐琅,是将无机玻璃质材料经熔融后凝于基体金属上,并与金属牢固地结合在一起的一种复合材料。珐琅彩,瓷器装饰手法之一,源于画珐琅技法。使用珐琅彩装饰手法的瓷器,即珐琅彩瓷(正式名称为"瓷胎画珐琅"),也常简称为珐琅彩。珐琅彩是将画珐琅技法移植到瓷胎上的一种釉上彩装饰手法,后人称"古月轩",国外称"蔷薇彩"。珐琅彩始创于清代康熙晚期,雍正时得到进一步发展。

(25) lead-arsenic 铅砷

(26) opaque white 玻璃白。粉彩瓷的彩绘方法一般是,先在高温烧成的白瓷上勾勒出图案的轮廓,然后用含砷的玻璃白打底,再根据深浅浓淡的不同需要,用干净的笔轻轻地将颜料施于这层玻璃白之上,将颜料洗开。

(27) armorial porcelain 徽章瓷、纹章瓷,是古代外销瓷的一种,把纹章,即欧洲诸国贵族、都市、团体等的特殊标志,印在瓷器上,故名。纹章瓷约始于明代。1974年英国出版的《中国纹章瓷》中,收集了约2000件纹章瓷。据瑞典统计,有300多家贵族曾到中国定烧过纹章瓷。

Exercises

Reading Comprehension Questions

Please answer the following questions according to the text.

(1) Who are the most famous supervisors in the imperial factories at Jingdezhen during the Qing dynasty?

(2) What are the features of famille verte enamels?

(3) What are the two decoration ways of famille verte enamels?

(4) What are the color features of Sang de boeuf glazes, peachbloom glaze and clair de lune glazes respectively?

(5) What are the features of famille rose enamels?

Translation

A. Please translate the following paragraphs into Chinese, paying special attention to ceramic terminology, context and the social background involved.

(1) Famille verte enamels painted on the biscuit. The decoration of porcelains painted with famille verte enamels on the biscuit is often set against a dominant, solid-color ground of black, yellow, green or occasionally, aubergine; the first two groups are frequently placed in the subcategories of famille noire and famille jaune. At their peak, these vessels are truly gorgeous, combining eloquent brushwork with well-mated colors and designs that are invariably right for the shape they grace. Three vases, representing the famille noire group, the famille jaune class, and the rare family with green grounds, are excellent examples of their respective types. They are the sort of porcelains that were greatly treasured by past generations in both this country and Europe.

(2) Famille rose. The signal accomplishment of the Yongzheng period was porcelain decorated in the famille rose palette of enamels. These opaque and semiopaque

enamels differ from the earlier overglaze polychrome enamels in two respects. First is the addition of a rose pink, from which this type of enameling takes its name. This pink, which was derived from colloidal gold, was used in a wide spectrum of tones from the palest blush of pink to deep ruby. Second—and perhaps even more important—a lead-arsenic, opaque white pigment was mixed with the colors to modify them, enabling the painter to achieve a range of color values for the first time. These new graduated tones allowed the artist to reproduce subtleties of shades and to model his drawing as the artist who paints with oils does. In addition, a variety of mixed tints was produced by combining colors, usually with the addition of the opaque white.

B. Please translate the following paragraph into English.

继康熙五彩瓷后，出现于康熙晚期的粉彩瓷至雍正朝达到历史上的最高水平。就像五彩瓷被称为"康熙彩"一样，粉彩瓷又被称为"雍正彩"。相对于浓烈明快、华贵浓艳的五彩瓷，粉彩瓷艳丽清雅，色调柔和，故五彩瓷被称为"硬彩"，粉彩瓷被称为"软彩"。粉彩瓷是在五彩瓷的基础上，由珐琅彩衍生出的一个新品种。它是在五彩所用的色料中掺以俗称"玻璃白"的氧化铝、硅、砷的化合物，利用其乳浊作用有意减弱色彩的浓艳程度，使用纸本绘画的渲染和没骨画法涂饰花纹，色阶繁多，色调温润柔和。

Unit 8 Introduction to Chinese Export Porcelain

    Chinese ceramics primarily for export can be traced back to the Tang dynasty, if not earlier. Export porcelain from China encompasses a vast array of porcelain items crafted predominately for shipment to Europe and later to North America from the 16th to the 20th century. These pieces differ from those intended for domestic use in China, both in terms of their shapes and their decorations, which were tailored to appeal to local tastes. The application of the term to wares made for non-western markets varies depending on the context. Chinese porcelain not only influenced the ceramic traditions of the countries it was exported to but also underwent influences from them in return.

Text

Porcelain Made for Exportation

Stephen W. Bushell

1 There had been a large quantity of Chinese porcelain exported to western countries from early **Mohammedan times**, when the Arabs first came to **Canton** by sea. Chinese fleets rode in the **Persian Gulf**, as related in their own annals of the ninth century, and confirmed by Mohammedan writers of the time. During the Yuan dynasty, when the same Mongolian house ruled Persia and China, the relations between the two countries became still more intimate, and there was constant traffic by land as well as by sea, for an account of which the celebrated *Travels of Marco Polo* may be consulted. In the Ming dynasty the overland route was barred by **the Mongolian Timur (the great Tamerlane)**, but Chinese ships continued to go west, touched at **Ceylon** and **Ormuz**, passed the **Strait of Bab-el-Mandeb** into the **Red Sea**, to land cargo at **Jidda**, the port of **Mecca**, and coasted the shore of Africa as far southward as **Zanzibar**. The voyages are described in detail in the Chinese annals of the reigns of Yongle (1403 – 1424 A. D.) and Xuande (1426 – 1435 A. D.). Early in the next century the Portuguese made their appearance in these seas, and from this time no more Chinese junks were seen in the Indian Ocean. The great mart was in the Persian Gulf, and any porcelain that reached Europe before the discovery of the voyage round the **Cape of Good Hope** would have come by caravan to **Cairo** or to **Aleppo**. Ancient Chinese porcelain has been found in the present day at many stations of the route that has been thus briefly sketched.

2 After the discovery of the route by the Cape of Good Hope, porcelain became better known in Europe. The Portuguese navigators appeared on the shores of **the Far East** in the beginning of the sixteenth century, and arrived at Canton in the year 1517, where they were at once admitted to trade. Japan was opened to them in 1542 by the shipwreck of a Portuguese vessel on the shore of the island of **Kyushu**, where they were well treated by the Japanese, and allowed to set up a trading establishment at **Nagasaki**. During the time that the Portuguese enjoyed the monopoly of the East

Indian trade they imported splendid collections of porcelain. The Dutch succeeded the Portuguese in the control of the trade with the Far East. **Van Neck** established a factory at **Batavia** in 1602, **the Dutch East India Company** was formed in the same year, and under its auspices vast quantities of porcelain were imported into Holland and the north of Europe.

3 The English East India Company, which was established in the reign of **Queen Elizabeth**, did not for a long period after its foundation succeed in opening a direct trade with China, being excluded by the Portuguese and Dutch. The port of **Gombron**, opposite to Ormuz, in the Persian Gulf, was for a long time the chief entrepôt of the British trade, and the earliest "China ware" introduced into England derived its name of "Gombron ware" from this place. In 1640 a factory was established at Canton, and direct trade has been carried on, with occasional interruptions, since that date.

4 With regard to the kinds of porcelain imported, the earliest porcelain imported was of **monochrome glaze**, principally **celadon** or white; **blue and white** followed. They seem to have been generally a selection from the ordinary contemporary productions of the private potters of Ching-te-chen. The work of the imperial manufactory could only have been exceptionally represented, as it is reserved for the service of the emperor. The private collections of Chinese connoisseurs were not ransacked, as they are in these later days, so that we can hardly expect to find any important examples of ancient ceramic art among the piles of dishes, plates, and tea services that were imported, as we gather from old bills of lading, by the hundred thousand.

5 It was in the reign of Kangxi (1662 – 1722) that porcelain seems to have been first made at Ching-te-chen in new forms and special designs for the European market. These were often executed after European models and designs taken there for the purpose by native agents from Canton. The earliest pieces with foreign designs were made for Persia and the Mohammedan market, and were decorated with scrolls of Arabic writing, generally texts from the **Koran**, the incorrect lettering of which, apart from the character of the floral designs with which they were associated,

betrayed the Chinese band. Next came Chinese copies of the old **Imari ware** of Japan, which were so perfectly executed during the reign of Kangxi that it would be sometimes difficult to distinguish the copy from the original were it not for the different quality and ring of the paste.

6   The name of "porcelaine des Indes" in France, of "India china" in England, was applied generally in the eighteenth century to the decorated Chinese porcelain which was imported in such large quantities, and eagerly sought after, until the time came when a similar material could be produced in Europe. Although the art of making **hard porcelain** was discovered in **Saxony** by **Böttger** in 1708, it was not till 1760 that it was made at **Sèvres**, and it hardly came into domestic use before the end of the eighteenth century. Meanwhile it was made and specially painted in China for exportation, and often from designs furnished by Europeans. A large proportion of it was evidently painted in Canton by Chinese artists, the porcelain being brought for the purpose overland from Ching-te-chen, glazed in the ordinary white state, with the addition perhaps of a few rings or outlines in **underglaze** blue defining the spaces intended to be filled in with colors. It is comparatively rare, however, in China, having been principally made for exportation and sent abroad at the time it was made.

7   Many of the services have on them the armorial bearings of the persons for whom they were made. The collection in **the British Museum** is very rich in this class of **"armorial china,"** including portions of services made for **Frederick the Great**, and for the royal families of Denmark and France, as well as many pieces with the arms of European families of rank, and of merchants who are known to have traded with China.

8   Some of the earlier pieces decorated with foreign designs were painted entirely in blue. The decoration was sometimes copied from European pictures brought to China for the purpose, so that we find in collections of Chinese porcelain sea views with Dutch vessels and **punch bowls** with pictures of English harvesting. The porcelain made to order for the European market, with which the Dutch inundated Europe for more than a hundred years, is generally overdecorated, in accordance with the

foreign taste. The vases of the same style and period being covered with richly dressed officials in their robes of office, have been sometimes classed apart under the title of "**mandarin porcelain**." This style is a favorite one with the Cantonese artist to the present day, when he is working for his foreign patron, although the native school of art, following always the canons of the old masters, disdains the modern costume of everyday life.

9 Among the objects made for Europe are found wash basins and ewers of elaborate form completely covered with floral brocaded **grounds** of diverse pattern, interrupted in the middle by a medallion with a **coat of arms**. The tea services which were imported consisted generally of a teapot with a hexagonal or octagonal tray, a pair of ovoid jars with covers as tea-caddies, a graceful cream jug with cover, one large bowl, a variable number of tea cups with or without handles, sometimes furnished with saucers, often without, and a plate or two for cakes, or a couple of saucer-shaped dishes.

10 There is one class of Chinese porcelain which has been dignified with the name of "**Jesuit china**," as it was supposed to have been made under the influence of the Jesuit missionaries. The pieces are usually painted in blue and white, and date from the earlier part of the reign of Kangxi. They are characterized by having the crucifix and other sacred symbols of the Roman Catholic faith introduced in the intervals of the decoration. The symbols in these specimens are penciled on the **paste** under the glaze, and must have been put on at the same time as the other part of the decoration, before the **firing**.

11 The decoration of oriental porcelain in Europe was first attempted in Holland. It was about 1700 that the potters are said to have discovered the secret of the preparation of a certain number of the colors of the **muffle stove**. These enamel colors, which were of the same class as those employed by the Chinese, were used not only for their own soft **faience**, but also in the decoration of hard porcelain imported from the Far East, being applied on white pieces, or on pieces spaced with a few blue lines, as prepared at Ching-te-chen for the artists of Canton, which were passed on to Europe for the purpose. Other pieces, in which the decoration appeared

to Dutch taste to be sparse, had the white ground filled in with various accessories and details of semi-oriental style, the result being a curious hybrid combination of colors as well as of styles.

Text Related Information

This text is adapted from chapter twenty-one of the book *Oriental Ceramic Art* (1897), written by Stephen W. Bushell, published by D. Appleton and Company, New York.

Selected Plates Related to the Text

Plate 38 Dish decorated with the Dutch ship Vryburg and the date 1756. Diameter: 22.8 cm. Qing dynasty.

Plate 39　Dish with the arms of England and the inscription "Engelandt". Diameter: 38.4 cm. Qing dynasty.

Plate 40　Two vases with portraits of Luther and Calvin. Height: 21.1 cm, 22 cm. Qing dynasty.

Unit 8　Introduction to Chinese Export Porcelain

Plate 41　Punch bowl with mandarin decoration. Diameter: 29.4 cm. Qing dynasty.

Aids to Comprehension

Notes

（1）Mohammedan times　穆罕默德时代。穆罕默德（约570—632），是阿拉伯宗教、社会和政治领袖,也是伊斯兰教的创始人。

（2）Canton　中国广州

（3）Persian Gulf　波斯湾,位于阿拉伯半岛与伊朗高原之间。

（4）*Travels of Marco Polo*　《马可·波罗游记》,记载了马可·波罗在1271年至1295年间从威尼斯出发至亚洲及从中国返回威尼斯的旅游经历,记述了亚洲及非洲多国的地理及人文风貌。该书很大程度上影响了欧洲人对东方的认识及探索。

（5）the Mongolian Timur（the great Tamerlane）　帖木儿帝国创建者,他经过一系列征服,形成东起北印度,西达小亚细亚,南濒阿拉伯海和波斯湾,北抵里海、咸海的帝国。

（6）Ceylon　锡兰,后改名为斯里兰卡,位于南亚次大陆以南印度洋上的岛国。

（7）Ormuz　同"Hormuz",霍尔木兹,位于霍尔木兹海峡的东北侧,现属伊朗。

（8）Strait of Bab-el-Mandeb　巴布—埃尔—曼德海峡,也称曼德海峡或曼达布海峡,是连接红海和亚丁湾的海峡,位于红海南端,也门和吉布提之间。其名在阿拉伯语中意为"泪之门"。

（9）Red Sea　红海,位于非洲东北部与阿拉伯半岛之间。其西北面通过苏伊士运河与地中海相连,南面通过曼德海峡与亚丁湾相连。

（10）Jidda　吉达,位于红海东岸,沙特阿拉伯麦加省的一个港口城市。

（11）Mecca　麦加,是沙特阿拉伯西部麦加省的首府。

（12）Zanzibar　桑给巴尔,位于东非坦桑尼亚联合共和国东部,由20多个岛屿组成。桑给巴尔一度为独立的岛屿,与阿拉伯世界有着悠久的贸易史,于1964年与坦噶尼喀（Tanganyika）合并组成坦桑尼亚。

（13）Cape of Good Hope　好望角,非洲西南端、大西洋沿岸的一个岩石岬角。

（14）Cairo　开罗,埃及的首都、第一大城市以及全国的经济、交通和文化中心。

该城市横跨尼罗河,是整个中东地区的政治、经济和商业中心。

(15) Aleppo 阿勒颇,叙利亚第一大城市

(16) the Far East 远东,西方国家对亚洲使用的地理概念,通常指东亚(含东北亚)、东南亚等离欧洲较远的地区。

(17) Kyushu 九州,位于日本西南部,为日本本土四岛之一。

(18) Nagasaki 长崎市,日本九州岛西部的都市,是日本西部重要的港湾都市。在江户时代的锁国时期,长崎是日本唯一的国际贸易港口,与荷兰、中国等有密切的交流。

(19) Van Neck: Jacob Corneliszoon van Neck 雅各布·科内利松·范·内克(1564—1638),荷兰海军军官、探险家。

(20) Batavia 巴达维亚,即现在的印度尼西亚首都雅加达。

(21) the Dutch East India Company 荷兰东印度公司,正名为联合东印度公司(The United East India Company),是荷兰历史上为向亚洲发展而成立的特许公司,于1602年成立,1799年解散。

(22) Queen Elizabeth: Elizabeth Ⅰ 伊丽莎白一世(1533—1603),1558年11月17日继位,1603年去世。

(23) Gombron 冈布龙,现在的阿巴斯港,位于伊朗南部,波斯湾霍尔木兹海峡沿岸的一座港口城市。

(24) Koran 《古兰经》

(25) Imari ware 伊万里瓷,是日本以有田为中心的肥前所产的瓷器的总称。尤其是在17世纪下半叶和18世纪上半叶之间,大量日本瓷器从日本伊万里港装船出口到欧洲。

(26) Saxony 萨克森州,全称为萨克森自由邦,是德国的一个内陆州,首府是德累斯顿。

(27) Böttger: Johann Friedrich Böttger 约翰·弗里德里希·博特格(1682—1719),德国炼金术士,通常被认为是在1708年发现硬质瓷器制造秘密的第一个欧洲人。

(28) Sèvres 全名为Manufacture nationale de Sèvres,塞夫尔国家制造厂。欧洲的主要瓷厂之一,位于法国上塞纳省的塞夫尔,前身是文森瓷厂。文森瓷厂成立于1740年,于1756年迁至塞夫尔。自1759年以来,它一直归法国

王室或政府所有。

(29) the British Museum　大英博物馆,一座位于英国伦敦布隆斯伯里的综合性博物馆,建立于1753年,1759年1月15日起正式对公众开放,是世界上首个国家博物馆,也是世界上规模最大的博物馆之一。馆内现有800多万件藏品。

(30) Frederick the Great: Frederick Ⅱ　腓特烈二世(1712—1786),史称腓特烈大帝,1740—1786年间在位的普鲁士国王。

(31) punch　潘趣酒,是17世纪初英国东印度公司的水手和雇员从印度带回英国的一种流行饮料,混合了多种元素,包括烈酒和果汁。从17世纪50年代起,盛放饮料的外销瓷碗开始流行起来。

(32) coat of arms　纹章,通常出现在纹章盾、外套或战袍上。中世纪骑士用它来辨认身份。盾徽由一个盾牌、支撑物、饰章及铭言组成。至今,它仍被作为识别个人、军队、教会、机关团体和公司企业的世袭或继承性标记。

Ceramic Terminology

(1) monochrome glaze: porcelain decorated with one single uniform colored glaze 单色釉,指用一种彩釉装饰的瓷器。

(2) celadon: wares glazed in the jade green celadon color and a type of transparent glaze that was first used on greenware, but later used on other porcelains. The term "celadon" was coined by European connoisseurs of the wares. The most commonly accepted theory is that the term first appeared in France in the 17th century and that it is named after the shepherd Celadon in Honoré d'Urfé's French pastoral romance *L'Astrée* (1627) wearing grayish-green ribbons and a matching coat. 青瓷,音译为雪拉同,指釉色翠绿的青瓷,也指一种透明釉,最初用于青瓷,后来用于其他瓷器。该词是由欧洲鉴赏家创造的。最普遍接受的理论是,这个词最早出现在17世纪的法国,来源于法国小说家奥诺雷·德·杜尔菲的小说《阿斯特蕾》(1627年)中身着灰绿色缎带和配套的外套的牧羊人。

(3) blue and white: also called blue and white porcelain, usually refers to ceramics

decorated with cobalt blue pigment on a white body, usually applied with a brush under the glaze. 青花瓷,指先用毛笔在瓷胎上画上钴蓝彩装饰,再施上表层透明釉的瓷器。

(4) hard porcelain: hard-paste porcelain 硬质瓷

(5) underglaze: a kind of decoration with the color painted on the body of a ceramic ware before glazing. 釉下彩,一种装饰工艺,先在器物胎体表面用彩料绘画装饰,然后再上一层釉料。

(6) armorial china: also called as armorial ware or heraldic china, refers to ceramics decorated with a coat of arms, either that of a family, or an institution or place. Armorials have been popular on European pottery from the Middle Ages and on porcelain from the 18th century. 纹章瓷器,是装饰有家庭纹章、机构纹章或者地区纹章等的陶瓷。从中世纪开始,纹章在欧洲陶器上就很受欢迎,18世纪的瓷器上都可以看到。

(7) punch bowl: a bowl, often large and wide, in which the drink punch is served. 潘趣碗,通常指又大又宽的碗,多用于盛放潘趣酒。

(8) mandarin porcelain: wares produced in China for export in the late 18th century with characteristics of the groups of figures in mandarin dress that appear in the decorative panels-painted mainly in gold, red, and rose pink and framed in underglaze blue. 人物瓷,18世纪晚期中国生产的出口瓷器。其特点是用一群身着清朝服饰的人物装饰——主要以金色、红色和玫瑰粉色绘制,并以釉下蓝为边框。

(9) ground 底色

(10) Jesuit china: also called as Jesuit ware and Jesuit porcelain 耶稣瓷

(11) paste: a mixture consisting mainly of clay and water that is used in making ceramic ware, especially a mixture of low plasticity based on kaolin for making porcelain. (制作陶、瓷器用的)湿黏土

(12) firing: the process of bringing clay and glazes up to a high temperature. The final aim is to heat the object to the point that the clay and glazes are "mature"—that is, that they have reached their optimal level of melting. This process is usually accomplished in two steps: bisque firing and glaze firing. 烧

制,将黏土和釉料加热到高温的过程。最终目的是将物体加热到黏土和釉料"成熟"的程度,即达到最佳的熔化水平。这一过程通常分为两个步骤:坯料烧制和釉料烧制。

(13) muffle stove: also called as muffle furnace or muffle oven, refers to a furnace in which the subject material is isolated from the fuel and all of the products of combustion, including gases and flying ash. 马弗炉,是指将被加热材料与燃料和所有燃烧产物(包括气体和飞灰)隔离开来的炉子。

(14) faience: earthenware decorated with opaque colored glazes　彩釉装饰的陶器;彩陶

Exercises

Reading Comprehensive Questions

Please answer the following questions according to the text.

(1) Please give a brief introduction to the history of exportation of Chinese porcelain.

(2) What does "Gombron ware" mean?

(3) What are the characteristics of Jesuit china?

(4) Why did the western countries continue to import Chinese porcelain even though they have acquired the technique of making hard porcelain during 18th century?

(5) How was the oriental porcelain further decorated in Europe?

Translation

A. Please translate the following paragraphs into Chinese, paying special attention to ceramic terminology, context and the social background involved.

(1) Among the objects made for Europe are found wash basins and ewers of elaborate form completely covered with floral brocaded grounds of diverse pattern, interrupted in the middle by a medallion with a coat of arms. The tea services which were imported consisted generally of a teapot with a hexagonal or octagonal tray, a pair of ovoid jars with covers as tea-caddies, a graceful cream jug with cover, one large bowl, a variable number of tea cups with or without handles, sometimes furnished with saucers, often without, and a plate or two for cakes, or a couple of saucer-shaped dishes.

(2) With regard to the kinds of porcelain imported, the earliest porcelain imported was of monochrome glaze, principally celadon or white; blue and white followed. They seem to have been generally a selection from the ordinary

contemporary productions of the private potters of Ching-te-chen. The work of the imperial manufactory could only have been exceptionally represented, as it is reserved for the service of the emperor. The private collections of Chinese connoisseurs were not ransacked, as they are in these later days, so that we can hardly expect to find any important examples of ancient ceramic art among the piles of dishes, plates, and tea services that were imported, as we gather from old bills of lading, by the hundred thousand.

B. Please translate the following paragraph into English.

据荷兰东印度公司的记载,每年仅巴达维亚一地运往欧洲的瓷器就达300万件之多,如果再加上其他国家及中国商人直接运往欧洲的瓷器,其数量之巨是可想而知的。明清时期的外销瓷主要是景德镇的青花瓷和釉上彩瓷。许多瓷器的装饰图案是依照外商从欧洲带来的样品由中国画工精心摹绘的,题材大致包括纹章(又称徽章)、人物故事、船舶及码头风景、动物、花卉、鱼草、博古等纹饰。在广州生产的"广彩瓷"即属于这一类。

Glossary

A

architectural pottery 建筑用陶器，简称"建陶"

armorial china 纹章瓷器

armorial porcelain 徽章瓷；纹章瓷

aubergine purple 茄紫

B

beater-and-pad technique 拍打印纹技法

biscuit 素胎；未施釉的烧制过的胎体

black-slipped pottery 覆盖黑色化妆土的陶器

blue and white 青花瓷

bright onion-green 葱翠

burnished pottery 抛光陶器

C

carbon-14 tests 碳十四测年法

celadon 青瓷

checker 方格图案；方格纹

cheng ting 挣钉

clair de lune glaze 月白釉

clay slice bonding 泥片围接法

clay 黏土

clear glaze 透明釉；清釉

colorant 着色剂

composition 构图

cong-style bottle 琮式瓶

cornaline 光玉髓；玛瑙

cushion pad 垫饼

D

dou 豆，中国先秦时期的食器和礼器

doucai 斗彩

double-lugged flat bottle 双系扁瓶

E

earthenware 陶器

earthworm trail pattern 蚯蚓走泥纹

egg white 卵白

enameling 彩饰；上光

end tile 檐梢瓦

engraving 刻花法

equidistant spinning pattern 等距旋纹

etch 凿刻

excavate 挖掘（古物）

F

face-upward firing 仰烧

faience 彩陶
boccaro 宜兴红陶
faience 彩釉装饰的陶器;彩陶
falangcai 珐琅彩
famille rose 粉彩;软彩
famille verte enamel(s) 硬彩;古彩;康熙五彩
fashion 制作
fencai 粉彩
figurine 小雕像
finish 修坯;利坯
finished 修坯工整的
firing 烧制
foot embellishing 修足
format album leaf 小册页
fuse the glaze 熔融釉料

G

glaze 釉
glazed pottery 上釉陶器
glaze-solvent 釉料助熔剂
grain 纹理;细颗粒
greenware 生坯
ground 底色
gu 觚

H

hard earthenware(s) 硬陶
high-fired glaze 高温釉
high-footed bowl 高足碗

hollowing out 镂空
hsiu hua 绣花;划花
hua hua 刻划花
Huiqing 回回青;回青

I

impress 压印
imprinted 印贴花法
incise 划;雕;刻;划花
inverted firing 覆烧

J

Jesuit china 耶稣瓷

K

kaolin 高岭土
kaolinite 高岭石;陶土

L

lacquered pottery 漆陶
Langyao 郎窑
lapis lazuli 青金石
lead-arsenic 铅砷
lead-glazed 施铅釉的
leather hard 半干状态
leys or slops jar(s) 渣斗
li tripod 鬲
lotus scroll 缠枝莲图案
low-fired 低温烧制的

M

Mandarin porcelain 人物瓷
material(s) 出土物
modeling 造型
mold 模子；模制成型法
monochrome glaze 单色釉
mortuary pottery 陪葬陶器；明器
motif 图案
mould 塑造；模制
muffle kiln 马弗窑；隔焰窑
muffle stove 马弗炉

O

opaque white 玻璃白
openwork 镂空雕刻
openwork 透雕
orange-red hue 橙红色色调

P

painted unglazed pottery 未上釉彩绘陶器
panel 开光
parrot green 鹦鹉绿
paste 湿黏土
peachbloom glaze 豇豆红釉；桃色釉
pedestal 器座
pingdeng 平等青；陂塘青
pi-se ware 秘色瓷
plain three-colors 素三彩
polish 磨光；精修

polychrome glazed ware(s) 颜色釉陶瓷；多彩釉陶瓷
poor plasticity 可塑性低
porcelain 瓷
Potangqing 陂塘青，亦称"平等青"
potted 胎体形状工整的；制作精良的
pottery 陶器；陶器、炻器和瓷器的总称
potter's wheel 陶车；拉坯车
potting 制作
punch bowl 潘趣碗；汁酒碗
purple rim and iron feet 紫口铁足

R

relief 浮雕
remains 遗存；遗迹。
reserve 防染本色区
rib 出筋
rolled rim 唇口
rough blue earth 大青土
round ware(s) 圆器
ruby red 红宝石

S

sang de boeuf glaze 牛血红釉
sculpting 塑像法
services 餐具套件
seven-lobed 七瓣形
sgraffito 剔花
shaped from wet clay 湿泥定型
short circular foot 矮圈足

slanting belly wall 斜腹壁
slip 化妆土;泥浆
solid-color ground 单色地
sprays of flowers 折枝花纹
stacking 堆贴;塑贴;堆塑
stamp 盖(章);印(某标记)
stoneware 炻器
string pattern 弦纹
strong reducing flame 强还原焰
sunflower rim 葵口

T

Tang sancai ware(s) 唐三彩
tear(s) 泪痕(瓷器在烧窑过程中形成的自然现象)
terra cotta 烧制后的红色黏土
texture 质地;纹理;肌理
the unbaked body 未烧制的坯体
thin-necked bottle 细颈瓶
throw 拉坯成型
translucent 半透明的
transparency 透明度
tripod stove 三足炉
Tzu 紫
tzu run 滋润

U

underglaze 釉底的;釉下彩

unglazed pottery 未上釉的陶器
unglazed rim 芒口

V

vermilion red 胭脂红
vitrify 玻璃化;使……呈玻璃状;烧结

W

wash 晕染;稀薄的液体
wax 蜡
white engobe 白色化妆土
wucai 五彩

X

Xing white ware(s) 邢窑白瓷

Y

yangcai 洋彩
yin hua 印花;贴花;模印贴花
yu 盂
Yue green ware(s) 越窑青瓷
yunlei pattern 云雷纹

Z

Zhe school artist 浙派艺术家
zun 尊

References

[1] 李知宴. 中国陶瓷艺术[M]. 北京:外文出版社,2010.

[2] 叶喆民. 中国陶瓷史[M]. 北京:生活·读书·新知三联书店,2011.

[3] BUSHELL S W. Oriental ceramic art[M]. New York: D. Appleton and Company,1897.

[4] FANG L L. The History of Chinese Ceramics[M]. Beijing: Foreign Language Teaching and Research Press,2013.

[5] HALL J H. Catalogue of late Yuan and Ming ceramics in the British Museum[M]. London: The British Museum Press,2001.

[6] HOBSON R L. Chinese pottery and porcelain[M]. New York: Funk and Wagnalls Company,1915.

[7] VALENSTEIN S G. A handbook of Chinese ceramics[M]. New York: The Metropolitan Museum of Art,1989.